INTEGRATED C
and
MICROPROCESSORS

CW01464317

APPLIED ELECTRICITY AND ELECTRONICS SERIES

Editor: P. HAMMOND, University of Southampton, UK

Pergamon Titles of Related Interest

BADEN FULLER	Engineering Field Theory
CORNILLIE	Microprocessors
HAMMOND	Electromagnetism for Engineers, 3rd Edition
HINDMARSH	Electrical Machines & Drives Worked Examples, 2nd Edition
HINDMARSH	Electrical Machines & Their Applications, 4th Edition
HOLLAND	Illustrated Dictionary of Microelectronics & Microcomputers
HOLLAND	Microcomputers & Their Interfacing
HOLLAND	Microcomputers for Process Control
HOWSON	Mathematics for Electronic Technology, 2nd Edition
KUFFEL & ZAENGL	High Voltage Engineering
RODDY	Introduction to Microelectronics, 2nd Edition
YORKE	Electric Circuit Theory

Pergamon Related Journals (*Sample copy gladly sent on request*)

Computers & Electrical Engineering

Electric Technology USSR

Microelectronics & Reliability

Robotics & Computer-Integrated Manufacturing

Solid State Electronics

INTEGRATED CIRCUITS
and
MICROPROCESSORS

R. C. HOLLAND
West Glamorgan Institute of Higher Education, UK

PERGAMON PRESS

OXFORD · NEW YORK · BEIJING · FRANKFURT
SÃO PAULO · SYDNEY · TOKYO · TORONTO

U.K.	Pergamon Press, Headington Hill Hall, Oxford OX3 0BW, England
U.S.A.	Pergamon Press, Maxwell House, Fairview Park, Elmsford, New York 10523, U.S.A.
PEOPLE'S REPUBLIC OF CHINA	Pergamon Press, Qianmen Hotel, Beijing, People's Republic of China
FEDERAL REPUBLIC OF GERMANY	Pergamon Press, Hammerweg 6, D-6242 Kronberg, Federal Republic of Germany
BRAZIL	Pergamon Editora, Rua Eça de Queiros, 346, CEP 04011, São Paulo, Brazil
AUSTRALIA	Pergamon Press Australia, P.O. Box 544, Potts Point, N.S.W. 2011, Australia
JAPAN	Pergamon Press, 8th Floor, Matsuoka Central Building, 1-7-1 Nishishinjuku, Shinjuku-ku, Tokyo 160, Japan
CANADA	Pergamon Press Canada, Suite 104, 150 Consumers Road, Willowdale, Ontario M2J 1P9, Canada

Copyright © 1986 Pergamon Books Ltd.

First edition 1986

Library of Congress Cataloging in Publication Data
Holland, R. C.
Integrated circuits and microprocessors
(Applied electricity and electronics)
Includes index
1. Integrated circuits. 2. Microprocessors.
I. Title   II. Series
TK7874.H62   1986   621.381'73   85-29647

British Library Cataloguing in Publication Data
Holland, R. C.
Integrated circuits and microprocessors
(Applied electricity and electronics)
1. Integrated circuits
I. Title   II. Series
621.381'73   TK7874

ISBN 0-08-033471-7 (Hardcover)
ISBN 0-08-033470-9 (Flexicover)

Printed in Great Britain by A. Wheaton & Co. Ltd., Exeter

Preface

During the 1970s and 1980s the great majority of electronic circuits have been re-designed so that they can be assembled into single silicon chips. This quantum leap in miniaturisation has also led to the ability to package enormously complicated circuits into single chips, e.g. the microprocessor. These "integrated circuits", or ICs, now totally dominate electronic system design, such that the modern circuit board appears as a bland array of black plastic packages, or "chips", each housing its silicon wafer circuit. This book is a concise learning package describing all the major IC types and their applications.

The three main categories of ICs are:

(a) digital circuits,
(b) analogue circuits,
(c) microprocessors and their support chips.

These and other miscellaneous devices are described in the text. Invariably books on electronic topics tend to describe just one of these categories of circuits, and therefore this text is an attempt to assemble all essential information to give the reader a reference for all types of circuits and chips he may encounter.

The book is aimed at the electronics student or practising engineer who has a rudimentary knowledge of electrical and electronic principles. It supports the principal electronic and microcomputer modules of most courses in electronic engineering. The material covers digital circuits (from gates to complex counter systems), analogue circuits ("op-amp" amplifiers and a wide variety of linear and non-linear applications) and microprocessors (the processing chip itself plus memory and input/output chips). Additional chapters cover conversion circuits of various types and fault finding, including descriptions of fault finding procedures and the latest forms of test equipment. The overall approach adopted is to introduce circuit principles, then details of the principal circuit building-blocks in chip form and finally application circuits.

The author wishes to thank his family and colleagues for their support during the preparation of this book.

Contents

CHAPTER 1

Integrated Circuits (ICs)

1.1 IC PHYSICAL APPEARANCE

An integrated circuit (or IC or "chip") can be defined as an electronic circuit in which several components are integrated into the same circuit package. It contrasts with a "discrete component circuit", in which each component, e.g. resistor, transistor, diode, is a separate device.

Several methods of packaging ICs have been applied since the first devices were produced in the 1960s. Virtually all modern ICs are produced in the DIL (dual-in-line) package, which is shown in Fig. 1.1. A silicon wafer, in which the circuit is deposited, connects to the interconnecting pins. Connections to other ICs are made along a circuit board using copper track conductors. The numbers of interconnecting pins are typically 8, 10, 12, 14, 16, 20, 24 or 40. Some recent microprocessors are packaged in 64-pin ICs.

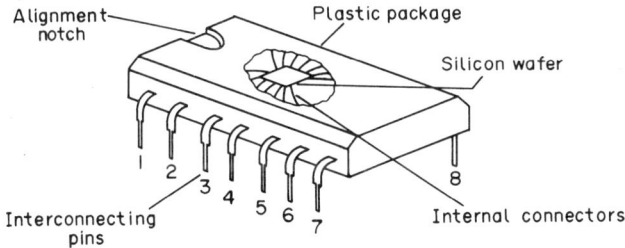

FIG. 1.1. Dual-in-line (DIL) package.

1.2 IC FABRICATION TECHNIQUES

The great majority of integrated circuits are constructed using the "planar" process. The stages in this process are:

(a) a crystal of silicon is grown in the shape of a cylinder, which is sliced to form circular wafers,

1

(a) Grow crystal
(silicon)

(b) Slice off wafer

(c) Deposit epitaxial
layer (into which
circuit will be
deposited)

UV Light

(d) Expose wafer to
UV light through a
photomask

(e) Etch away
unexposed areas

(f) Deposit impurity
in etched areas

Repeat (d) to (f) for several stages of deposition

(g) Scribe and break
wafer into several
individual chips

(h) Encapsulate chip into
plastic package

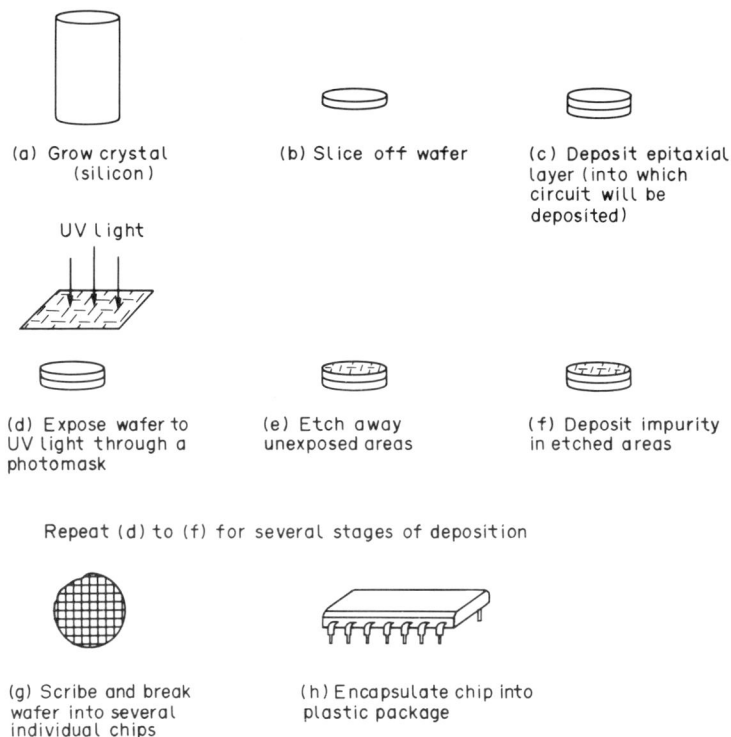

FIG. 1.2. Fabrication process of integrated circuit.

(b) a series of photomask and vapour diffusion processes is carried out on each wafer in order to implant regions within the silicon with impurity elements, e.g. boron and phosphorus, to fabricate transistors and other components,

(c) the wafer is scribed and broken into typically several hundred individual and identical circuits before each is encapsulated into its final plastic package.

This procedure is illustrated in Fig. 1.2.

A cross-section of a silicon wafer that supports a single transistor is shown in Fig. 1.3. Figure 1.4 shows the circuit symbol for this conventional "bipolar" transistor (so called because both positive and negative charge carriers exist). Other components, e.g. resistors and diodes, can be diffused into the same wafer to construct complete circuits. Metallic interconnections are diffused across the top of the silicon oxide layer.

"Unipolar" transistors (so called because only a single type of charge

Collector lead   Base lead   Emitter lead

Silicon oxide (insulator)

Epitaxial n layer into which p (transistor base) and n (transistor emitter) are diffused

n type (boron diffusion)

p type (phosphorus diffusion)

Silicon substrate

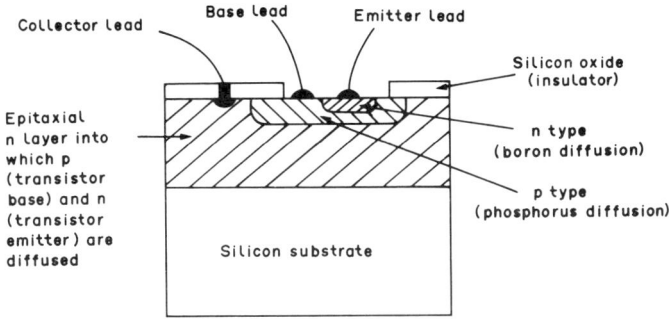

Fig. 1.3. Planar transistor (cross-section of silicon wafer).

carrier exists) can also be fabricated in integrated circuit form, and they have caused the huge jump in circuit micro-miniaturisation that has led to the development of the microprocessor and the single-chip microcomputer. The unipolar transistor is better known by its name of FET (field effect transistor), and the construction of an insulated gate FET (IGFET) is illustrated in Fig. 1.5. When the gate voltage goes more positive, the channel conductivity between drain and source increases, i.e. the channel current is "enhanced". The gate input connection is electrically insulated from the n channel by a layer of silicon dioxide, and this contributes to the name MOS (Metal Oxide Silicon) for this type of transistor fabrication. The circuit symbol for the device is shown in Fig. 1.6.

Unipolar transistor circuits, which are called MOS and CMOS (Complementary MOS), can be constructed with far higher packing density than bipolar transistor circuits, which are principally constructed using TTL (Transistor Transistor Logic) technology. Whilst these circuit families are described later in this book (TTL in chapters 4 and 5, MOS and CMOS in

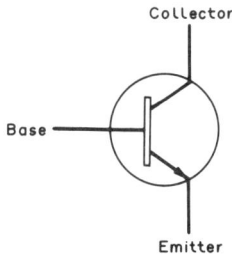

Collector

Base

Emitter

Fig. 1.4. Circuit symbol for conventional unipolar transistor.

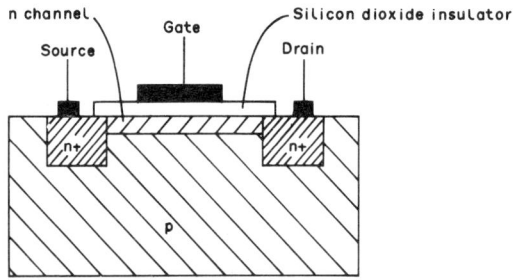

FIG. 1.5. Insulated gate FET (cross-section of silicon wafer).

chapters 10, 11 and 12), it is worthwhile here emphasising the principal circuit families and their applications:

(a) Bipolar circuits, principally TTL and its variation ECL (Emitter Coupled Logic)—used for fast switching logic gates, registers and counters

(b) Unipolar circuits, MOS and CMOS—used for high-density micro-processor and memory chips.

The great majority of integrated circuits are "digital" in operation, i.e. they process only discrete signal levels. However, some "analogue" circuits (e.g. small signal amplifiers), which process signals that are continuous, are also packaged in integrated circuit form, and they are described in Chapters 7 and 8.

1.3 CIRCUIT BOARDS

The full assembly of a modern electronic circuit is made invariably using a base board of epoxy glass construction supporting a range of ICs and other

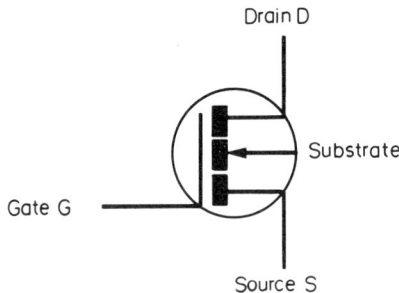

FIG. 1.6. Circuit symbol for FET (n channel).

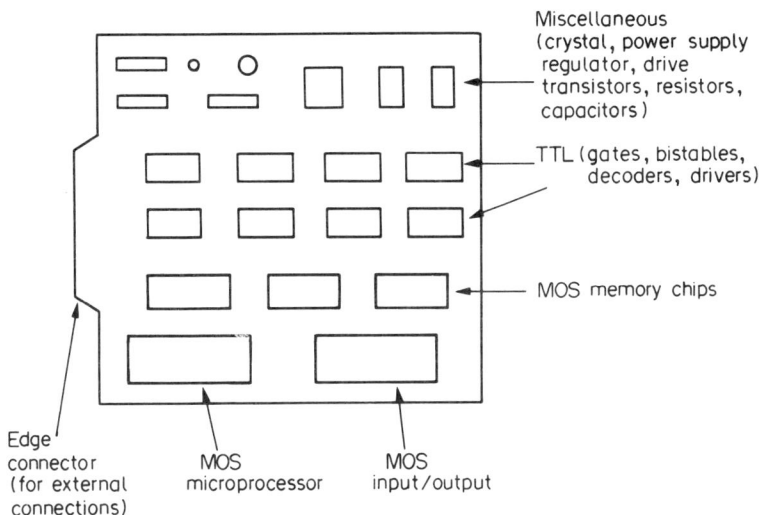

FIG. 1.7. Typical circuit board.

components, as illustrated in Fig. 1.7. The ICs are inserted into drilled holes in the board, and the reverse side of the board supports copper tracks which carry circuit interconnections. IC pins are soldered to these tracks. The method of laying the copper tracks involves preparing a photomask of the interconnection network required, and then using this with a photoresist and etching process to remove the majority of the copper plating on a fully-plated source board. Board assembly is often then carried out using automatic IC insertion and automatic soldering equipment, particularly in high-volume production applications.

ICs that may need to be replaced, e.g. memory chips, or are more prone to failure, are often mounted in IC sockets so that no unsoldering is required when they are removed.

Larger assemblies may require several circuit boards, which are mounted into a "back-plane" which provides inter-board connections. One board in such a system typically performs a power supply function, i.e. it converts ac mains into dc supply voltages (predominantly +5 V and 0 V) to energise the chips in the other circuit boards.

Integrated circuits generally fall into three categories:

(a) TTL digital chips, or their CMOS equivalents
(b) MOS microprocessor (and allied functions) chips, or their CMOS equivalents
(c) auxiliary chips, e.g. analogue amplifiers.

In a digital circuit, e.g. a microcomputer circuit board, a general rule-of-thumb is that if a chip possesses more than 20 pins it is probably MOS, but if it possesses less than 20 pins it is probably TTL.

BIBLIOGRAPHY

1. *Digital Integrated Circuits and Computers*, Barry Woollard, McGraw-Hill, 1978.
2. *Microchip Technology*, Charles Kerridge, John Wiley, 1983.

EXERCISES

1. What is a DIL?
2. What is the purpose of an "epitaxial layer" in the construction of a silicon chip?
3. Justify the description "insulated gate" in an IGFET.
4. State an advantage and a disadvantage of unipolar circuits compared with bipolar circuits.
5. To which circuit family do microprocessors normally belong—TTL, MOS or CMOS?

CHAPTER 2

Logic Gates

2.1 INTRODUCTION

Electronic circuits and computers process data in the form of two-state signals, e.g.

voltage, or no voltage
short-circuit, or open-circuit
on, or off
0, or 1

Circuits which combine such signals are called "gates". Alternative circuits can memorise or otherwise process these signals. This chapter examines the operation of gates, and later chapters describe how gates are employed in complete electronic systems, including computers.

Two-state signals can be described as "binary digital" signals. They are digital because they can take only a limited number of states, unlike analogue signals which can take any value over a continuous range. They are binary because they can take one of only two states. Frequently a signal of this type is termed a "bit" (*bi*nary digi*t*). The most common method of representing these two states in electronic circuits is:

logic 0 = 0 V
logic 1 = +V (most commonly +5 V)

These signals are reversed if "inverse logic" is used, i.e. logic 0 = +V and logic 1 = 0 V. Alternative terminology is "positive logic" (normal logic) and "negative logic" (inverse logic).

Standard gate circuits, as well as bistables (memory circuits), counters and other circuits to be described in Chapters 4 and 5, are available in TTL form in the "7400 series". This range of ICs has dominated digital circuits since the early 1970s. Although augmented by MOS circuits, which offer much more complex circuit functions in chip form, the 7400 range is still applied for the standard gating and switching functions that are described in this chapter and in chapters 4 and 5. A diagrammatic list of the principal chips in the 7400 range is given in Appendix A.

2.2 AND GATE

Figure 2.1 shows the circuit symbol for an AND gate which combines two single-bit signals. The output is set to 1 only if both inputs are 1. All possible signal states are summarised in the "truth table". The AND function is denoted by the dot ("period") logic symbol between the two input signal identities.

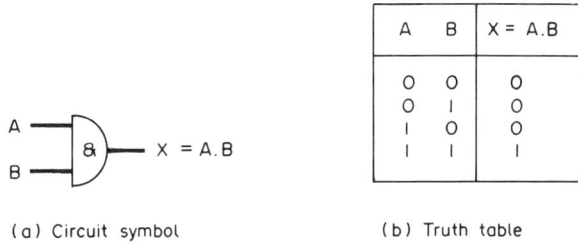

A	B	X = A.B
0	0	0
0	1	0
1	0	0
1	1	1

(a) Circuit symbol   (b) Truth table

FIG. 2.1. AND gate.

The AND gate is available in the TTL 7400 series of ICs as the 7408, as shown in Fig. 2.2. Four gates are built onto the same chip, which is termed a "quadruple 2-input AND gate. The dc supply voltage of +5 V is connected using the V_{CC} and GND pins.

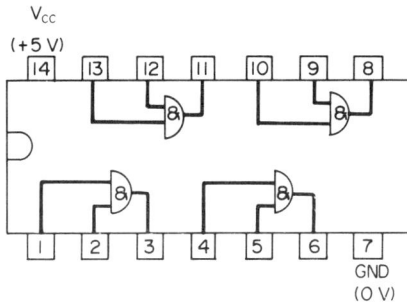

FIG. 2.2. Quadruple 2-input AND gate (7408).

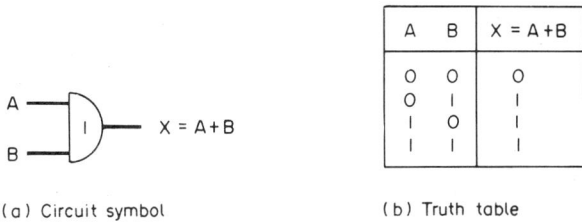

A	B	X = A+B
O	O	O
O	I	I
I	O	I
I	I	I

(a) Circuit symbol   (b) Truth table

FIG. 2.3. OR gate.

2.3 OR GATE

Figure 2.3 shows the circuit symbol for an OR gate. The output is set to 1 if either (or both) of the inputs is 1. The OR function is denoted by the + logic symbol.

The OR gate is available in the 7400 series as the 7432, as shown in Fig. 2.4.

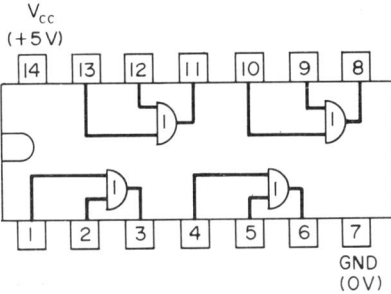

FIG. 2.4. Quadruple 2-input OR gate (7432).

2.4 NAND GATE

A NAND (or NOT AND) gate is simply an AND gate with the output signal inverted, as shown in Fig. 2.5. The inversion process is indicated by the use of a bubble at the output connection. Additionally the logic inversion process is indicated by the bar above the whole logic expression.

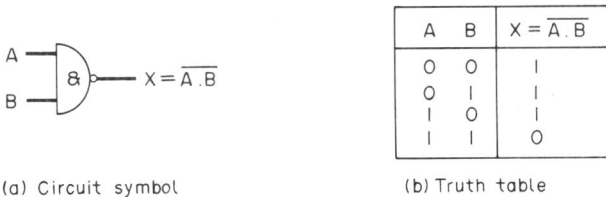

A	B	$X = \overline{A.B}$
O	O	I
O	I	I
I	O	I
I	I	O

(a) Circuit symbol   (b) Truth table

FIG. 2.5. NAND gate.

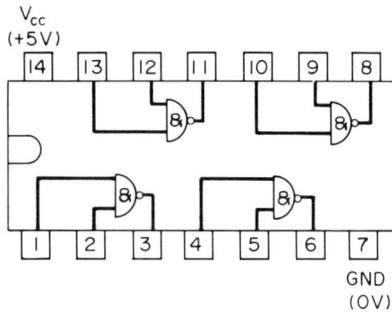

FIG. 2.6. Quadruple 2-input NAND gate (7400).

The NAND gate is available in the 7400 series as the 7400, as shown in Fig. 2.6.

2.5 NOR GATE

A NOR (or NOT OR) gate is an OR gate with the output signal inverted, as shown in Fig. 2.7. Figure 2.8 shows the 7402 TTL package for four such gates.

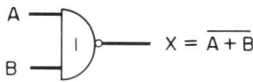

A	B	$X = \overline{A+B}$
O	O	I
O	I	O
I	O	O
I	I	O

$X = \overline{A + B}$

(a) Circuit symbol   (b) Truth table

FIG. 2.7. NOR gate.

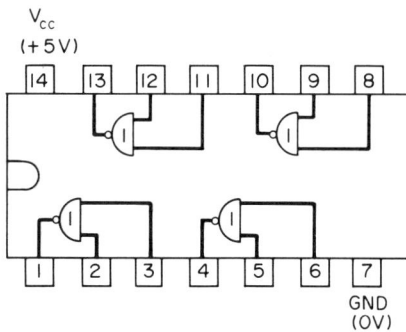

FIG. 2.8. Quadruple 2-input NOR gate (7402).

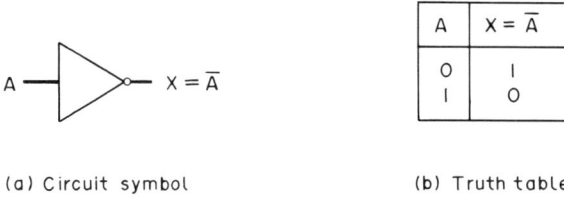

A	X = Ā
O	I
I	O

(a) Circuit symbol   (b) Truth table

FIG. 2.9. Inverter.

2.6 INVERTER

An inverter, or NOT gate, simply inverts the single-bit input signal, as shown in Fig. 2.9. It is available in the 7400 range as the 7404, which supports six such inverters, as shown in Fig. 2.10.

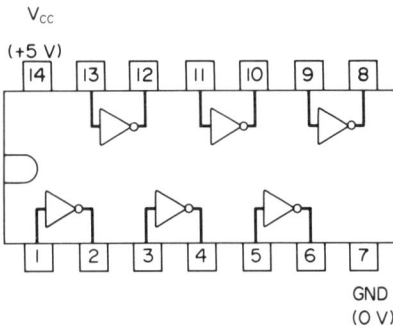

FIG. 2.10. Hex inverter (7404).

Frequently it is inefficient to include an inverter chip in a circuit if there are unused NAND or NOR gates available, e.g. if only three of the four gates on a quadruple NAND gate chip are used. In this case the invert function can be generated as shown in Fig. 2.11.

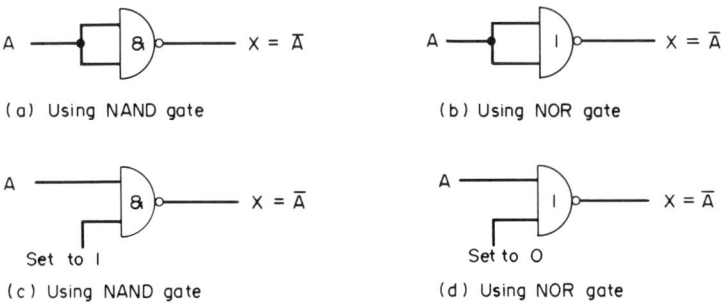

(a) Using NAND gate

(b) Using NOR gate

(c) Using NAND gate

(d) Using NOR gate

FIG. 2.11. Generation of single-bit invert function using NAND and NOR gates.

2.7 BOOLEAN ALGEBRA

In the nineteeth century George Boole devised a set of algebraic rules and theorems that can be applied to logic systems. His rules allow logic circuitry, which processes two-state signals, to be expressed symbolically. This in turn leads to the ability of a designer to reduce circuit complexity.

The Boolean symbolism for the AND, OR, NAND, NOR and NOT logic functions has already been introduced in this chapter. The following Boolean rules allow the operation of logic circuits to be expressed symbolically and to be better understood:

Using AND function:

$$A.1 = A$$
$$A.0 = 0$$
$$A.A = A$$
$$A.\overline{A} = 0$$
$$A.B = B.A$$

Using OR function:

$$A+1 = 1$$
$$A+0 = A$$
$$A+A = A$$
$$A+\overline{A} = 1$$
$$A+B = B+A$$

Also $A = \overline{\overline{A}}$

All of these relationships can be confirmed by truth tables.

Two additional rules, devised by de Morgan, further enable circuits to be simplified. These rules are:

$$\left. \begin{array}{l} \overline{A.B} = \overline{A}+\overline{B} \\ \overline{A+B} = \overline{A}.\overline{B} \end{array} \right\} \text{ de Morgan's rules}$$

The significance of de Morgan's rules is that complete logic systems can be designed using only NAND gates or NOR gates. Figure 2.12 shows how an AND function can be generated using NOR gates, whilst Fig. 2.13 shows how an OR function can be generated using NAND gates. Additionally complex circuits can be simplified by using these rules. Clearly an AND function can be generated using NAND gates by employing a NAND gate followed by a NAND gate used as an inverter. Similarly an OR function is generated by using a NOR gate followed by a NOR gate used as an inverter.

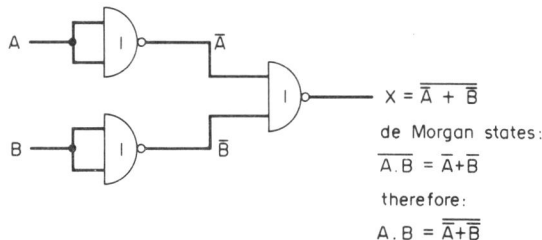

$$X = \overline{A} + \overline{B}$$

de Morgan states:

$$\overline{A.B} = \overline{A}+\overline{B}$$

therefore:

$$A.B = \overline{\overline{A}+\overline{B}}$$

Fig. 2.12. Generation of AND using NOR gates.

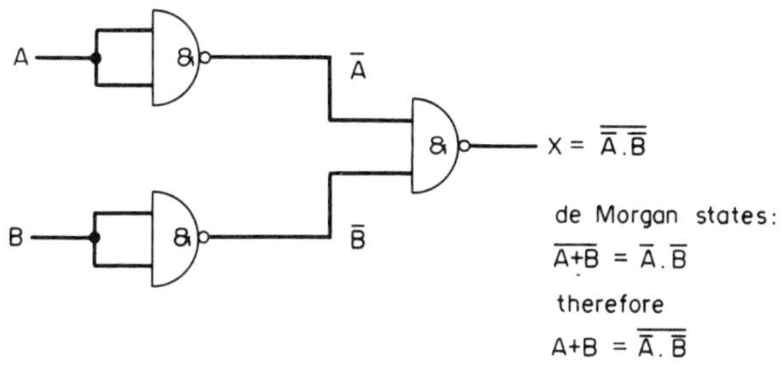

de Morgan states:

$$\overline{A+B} = \overline{A}.\overline{B}$$

therefore

$$A+B = \overline{\overline{A}.\overline{B}}$$

FIG. 2.13. Generation of OR using NAND gates.

2.8 LOGIC GATE CIRCUITS

Frequently gates with more than two inputs are required. Several chips within the 7400 series offer multiple inputs; these normally possess 3 or 4 inputs. The reader is referred to Appendix A for the full list of available devices.

Logic circuit diagrams frequently show different circuit symbols for the standard range of gates. Figure 2.14 illustrates the different symbols which are widely used. Although the second of the two sets of British symbols is more recent than the first, the first is still more commonly used and is applied in this book.

A sample logic circuit is shown in Fig. 2.15. Notice that the OR gate has the legend U3. This is the IC number in this particular circuit. Sometimes chips are labelled IC1, IC2, etc., in place of U1, U2, etc. This standard augments the labelling system for transistors (Q1, Q2, etc.), resistors (R1, R2, etc.), capacitors (C1, C2, etc.) and others. Notice that the two inverters

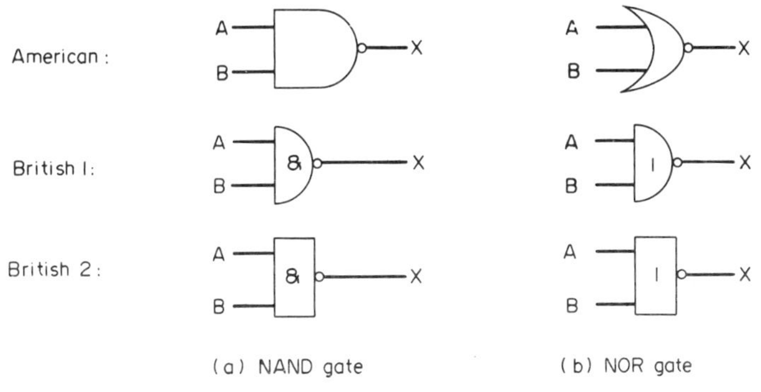

FIG. 2.14. Different circuit gate symbols.

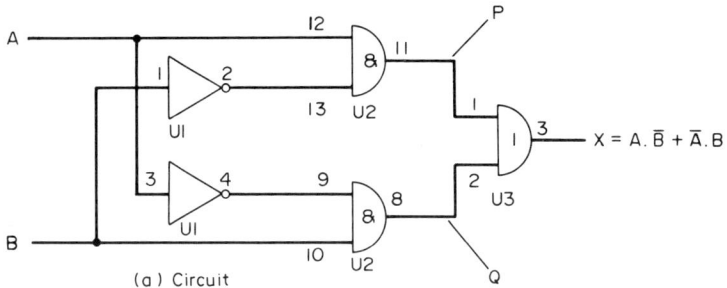

(a) Circuit

A	B	P	Q	X
0	0	0	0	0
0	1	0	1	1
1	0	1	0	1
1	1	0	0	0

(b) Truth table

FIG. 2.15. Typical logic circuit.

share the same legend U1. This is because they share the same IC package. For example, the inverters may form two out of six inverters in a 7404, and the two AND gates may form two out of four AND gates in a 7408—both chips were described earlier in this chapter. The pin numbers for each gate are also shown in the diagram.

An alternative method of indicating that a circuit function is performed using only part of an IC package is as follows:

$\frac{1}{4}$ 7408   for each of the AND gates

$\frac{1}{6}$ 7404   for each inverter.

This particular circuit has a special function. The truth table shows the intermediate signal levels in the circuit, and these can be expressed as follows:

$$P = A.\overline{B} \quad \text{and} \quad Q = \overline{A}.B$$

The final output signal can be seen to be set to 1 only when the inputs are different (0 and 1, or 1 and 0). For this reason the circuit can be used as a "comparator". It possesses the alternative logic names of EXCLUSIVE OR and NOT EQUIVALENT.

BIBLIOGRAPHY

1. *Digital Electronics: Fundamental Concepts and Applications*, Christopher E. Strangio, Prentice/Hall, 1980.

2. *Using Digital and Analogue Integrated Circuits*, L. W. Shacklette and H. A. Ashworth, Prentice/Hall, 1978.

EXERCISES

1. What voltage level normally represents logic 1 in an "inverse logic" system?
2. How many interconnecting pins would you expect to find in a DIL package which offers a triple 3-input AND gate? Check your answer with Appendix A.
3. What is the logic state of the output of a NOR gate when both inputs are at logic 1?
4. Simplify the Boolean expression:

$$A + A.1$$

5. Sketch a logic circuit using only NOR gates to generate the following logic function:

$$A + (B.C)$$

6. Repeat 5, using only NAND gates.
7. Express the output X of the circuit below in terms of its three inputs using a Boolean expression. What is the logic state of the output when A, B and C are at logic 0, 0 and 1 respectively. (Hint: draw out a truth table, including the state of the intermediate output of the OR gate.)

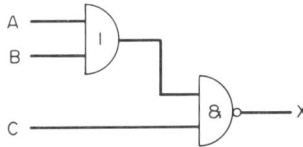

CHAPTER 3

Logic Families

3.1 INTRODUCTION

Several electrical and electronic circuits have been applied over the last 40 years to perform logic and gating functions. The earliest form of electronic logic utilised thermionic valves. Clearly the latest circuits employ integrated circuits, which are normally TTL. In this chapter we will examine the basic circuit families of TTL and MOS, and variations of these families, but in order to assist an understanding of the operation of the TTL range of devices a description of its predecessor, DTL, is presented. Also switch circuit logic is described for two reasons:

(a) it is a commonly applied system of logic that is used when an electronic solution (using power supplies and circuit boards) is unwieldy or inefficient,

(b) it assists an understanding of more complex electronic logic.

A brief primer on the use of the diode and the transistor as switching elements may be useful before the operation of the different circuit families is examined. Firstly, the action of a diode is illustrated in Fig. 3.1. When the diode is forward biased, by applying a more positive voltage to its anode compared with its cathode, the diode conducts and passes current. When the voltage polarity across the device is reversed, the diode does not conduct and no current flows.

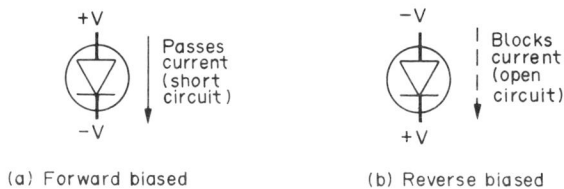

+V
Passes
current
(short
circuit)
−V

(a) Forward biased

−V
Blocks
current
(open
circuit)
+V

(b) Reverse biased

FIG. 3.1. Switching action of diode.

+V

Connect
input
to +V

Output = logic 0 (0V)

C

Passes current
(short circuit)

B

E

0V

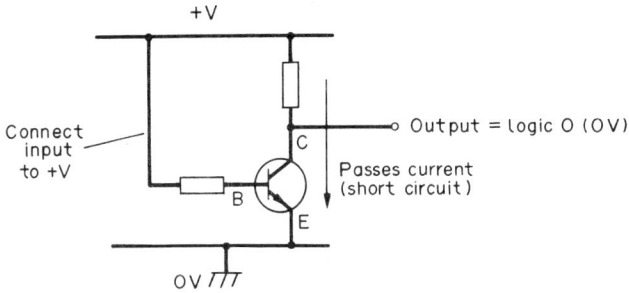

(a) Input at Logic 1 (+V) — transistor switches on

C = Collector
E = Emitter
B = Base

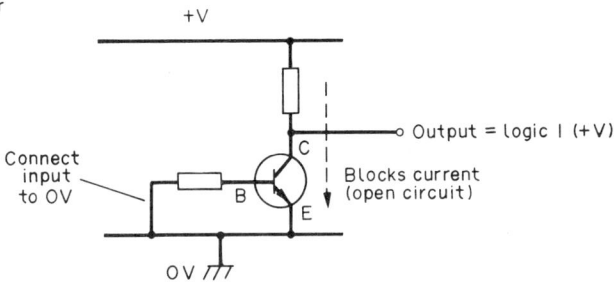

+V

Connect
input
to 0V

Output = logic 1 (+V)

C

Blocks current
(open circuit)

B

E

0V

(b) Input at Logic 0 (0V) — transistor switches off

FIG. 3.2. Switching action of transistor.

The switching action of the transistor is illustrated in Fig. 3.2. When the input is connected to logic 1, the transistor switches on (an approximate short-circuit exists between the C and E connections) and the output voltage is set at logic 0. When the input is set at logic 0, the transistor switches off (an open-circuit exists between the C and E connections) and the output voltage is set at logic 1. Therefore the transistor performs an inversion process as well as providing a switching function.

3.2 SWITCH CIRCUIT LOGIC

This form of logic employs simple electric switches to perform the standard gating functions, as shown in Fig. 3.3. A and B can be set as follows:

(a) manually, e.g. toggle switch, latching pushbutton,
(b) automatically, e.g. relay contact, limit switch, thermostat.

A   B

I ————o/ o———o/ o———— X = A.B

(a) AND gate

A

I ————o/ o——— X = A + B
      B

(b) OR gate

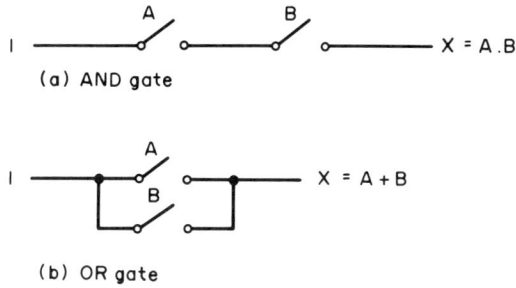

FIG. 3.3. Switch logic gates.

The connection of switches, or "contact closures", in series produces an AND function. Connecting them in parallel produces an OR function. Complete logic systems can be constructed using these simple rules, and such systems are commonly applied with the control of electrical machinery. A simple domestic example is the control of a central heating pump, as shown in Fig. 3.4. There are four control signals, or contact closures, in the system, and using the signal identities shown in the diagram:

$$X \text{ (to activate pump)} = O.R.(M+C)$$

Thus the pump is activated if:

the on/off switch (O) is set to on
AND
the room thermostat (R) is set (room temperature low)
AND
manual over-ride (M) is set *OR* clock contact (C) is set (on time clock).

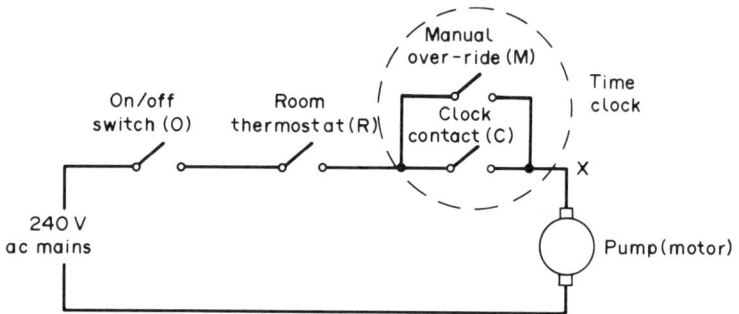

Manual
over-ride (M)

On/off   Room   Time
switch (O)   thermostat(R)   Clock   clock
                              contact (C)

240 V
ac mains   Pump(motor)

FIG. 3.4. Example of switch logic—central heating pump circuit.

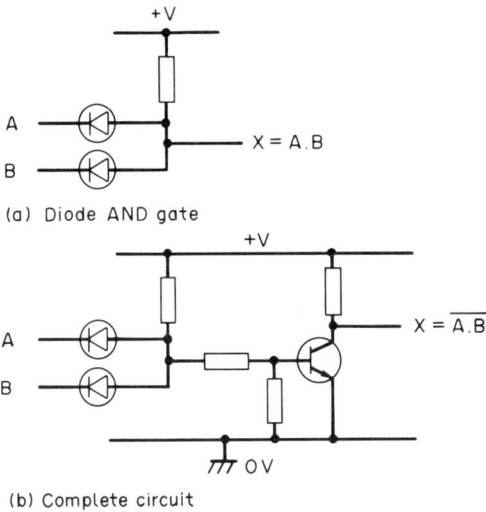

(a) Diode AND gate

(b) Complete circuit

Fig. 3.5. DTL NAND gate.

3.3 DTL (DIODE TRANSISTOR LOGIC)

DTL has been almost totally superseded by later forms of electronic logic families. However it is worthy of consideration here because an understanding of circuit operation will assist the reader when the other families are examined.

Consider the circuit of Fig. 3.5. In (a) a simple diode AND gate produces a logic 1 (+V) output only when both inputs are set to 1 (connected to +V)—refer to section 2.2 to confirm the truth table for an AND gate. If either (or both) of the inputs is held at logic 0 (connected to 0 V), then a short-circuit between output X and 0 V exists through the conducting diode/diodes, and X is at logic 0. When the output of the diode AND gate is passed through an inverting transistor in the complete circuit in (b), a NAND function is created.

3.4 TTL (TRANSISTOR TRANSISTOR LOGIC)

TTL circuits dominate gating circuit applications. Despite the advent of MOS technology for microprocessors and their support chips, TTL is still the most common IC family.

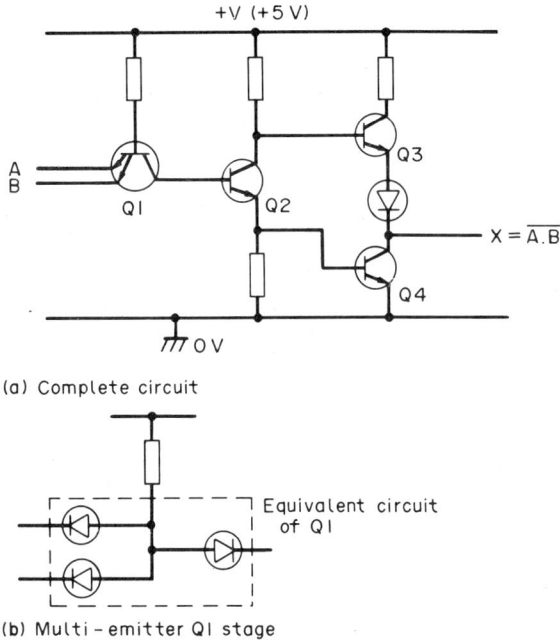

(a) Complete circuit

(b) Multi-emitter QI stage

Fig. 3.6. TTL NAND gate.

The basic TTL NAND gate is shown in Fig. 3.6. The input transistor Q1 is a multi-emitter device—it would possess three emitters if the gate was a triple-input NAND gate. Its equivalent circuit is simply a diode AND gate, as just described in section 3.3. The overall circuit produces a logic 0 (0 V) output only when both inputs are at logic 1 (+V). In this state current flows out of Q1 collector, which is at +V, into the base connection of Q2. Q2 turns on, so that the emitter current from Q2 feeds base current to Q4, which turns on. The voltage on Q2 collector is low enough to turn Q3 off. Therefore the output X is at 0 V (logic 0).

If either (or both) of the inputs is at logic 0, current flows out of the emitter/emitters of Q1. There is no base current to Q2, which turns off, and this in turn causes Q4 to turn off. The high voltage on Q2 collector turns Q3 on, and the output X is at +V (logic 1).

The output circuit (Q3 and Q4) is often described as a "totem pole" output, and it gives a low impedance drive in both high and low output states. This provides high speed operation with the ability to drive capacitive loads.

TTL offers two advantages over DTL: the multi-emitter transistor

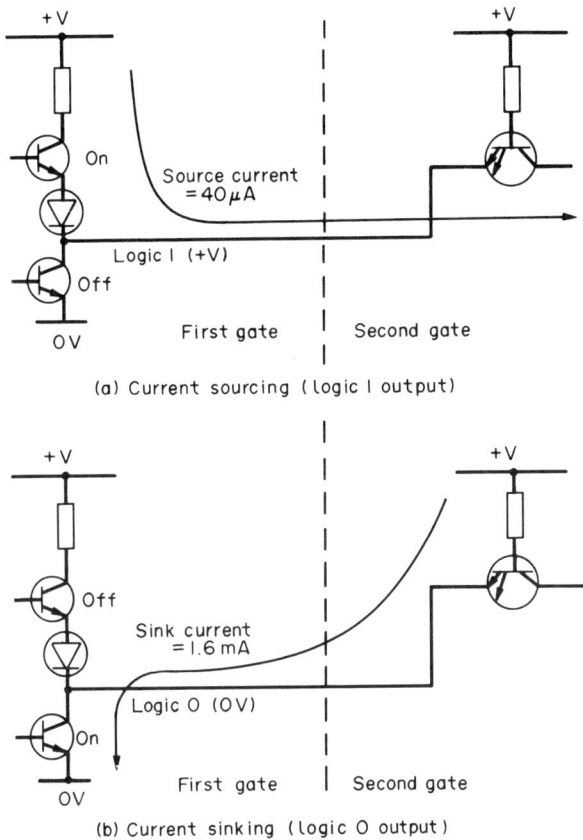

(a) Current sourcing (logic I output)

(b) Current sinking (logic O output)

FIG. 3.7. TTL current sourcing and sinking.

requires a smaller fabrication area than the equivalent number of diodes, and its switching speed is approximately twice as fast.

When TTL gates are connected together the first, or driving, gate is said to "source" or "sink" current to the second gate, as illustrated in Fig. 3.7. When the output from the first gate is logic 1, current flows from the first gate into the second gate. Current flows in the opposite direction for logic 0. TTL gates are normally designed to possess a "fan-out" of 10, i.e. the maximum number of gates that a single gate can drive is 10, as follows:

Single gate source current $= 40\,\mu A$; Maximum $= 400\,\mu A$; Fan-out $= 10$.
Single gate sink current $= 1.6\,mA$; Maximum $= 16\,mA$; Fan-out $= 10$.

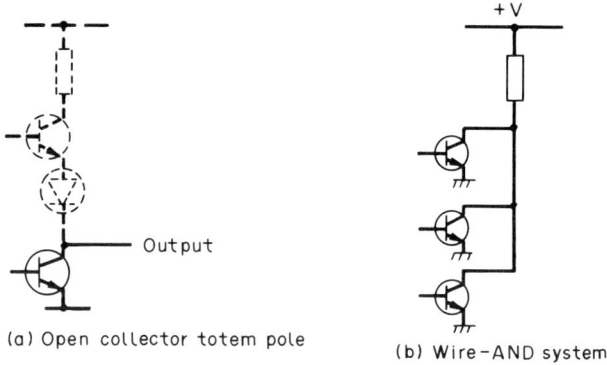

(a) Open collector totem pole

(b) Wire-AND system

Fig. 3.8. TTL open collector gates.

A variation of the totem pole output circuit is the "open collector" configuration, shown in Fig. 3.8(a), in which only the lower transistor Q4 is used. An external load to +V must be connected. In Fig. 3.8(b) several open collector gates are wired together to perform an AND function using only one resistor. Thus the use of open collector gates can reduce the total number of gates in the system when AND gates are required.

The standard range of TTL integrated circuits is the SN7400 series—the SN prefix stands for "Semiconductor Network", and is often omitted. A rarely used alternative is the SN5400 series, which offers a much wider operating temperature range and is intended for military applications. Variations of the SN7400 series are:

(a) SN74L00   — L indicates low power (but slower switching speed than SN7400),

(b) SN74S00   — S indicates Schottky (a diode is placed between the base and collector of each transistor in order to increase the switching speed),

(c) SN74LS00   — low power Schottky (factor of five reduction in power compared with SN7400 series),

(d) SN74ALS00 — advanced low power Schottky (twice the speed of ordinary Schottky with half the power of low power Schottky).

The SN74LS00 series is typically only 10% more expensive than the SN7400 series, and so is normally preferred. The SN74ALS00 series is approximately twice the cost of SN7400. The last two (or three) digits, e.g. 15 in SN74LS15, indicate the device function (triple 3-input NAND gate in this case). A full list of the TTL 7400 range of devices is given in Appendix A, and several of these chip functions will be examined in chapters 4 and 5.

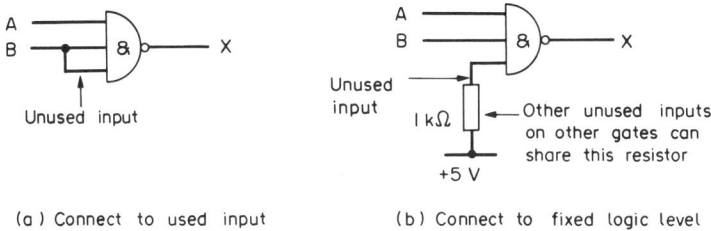

(a) Connect to used input   (b) Connect to fixed logic level

Fig. 3.9. Connection of unused TTL inputs.

The following points should be observed when TTL circuits are designed:

(1) The acceptable supply voltage range is restricted to 4.75 V to 5.25 V, and so good dc supply voltage regulation should be maintained.
(2) Any unconnected input acts as a logic 1; also it can pick up electrical noise. One of the circuit arrangements shown in Fig. 3.9 should therefore be used when a gate input is not utilised.

When testing or fault-finding TTL circuits, do not expect +5 V and 0 V voltage levels for logic 1 and logic 0. The following voltage ranges are acceptable within a typical specification:

$$\text{logic } 1\text{—}+2 \text{ V to } +5 \text{ V}$$
$$\text{logic } 0\text{—}0 \text{ V to } +0.8 \text{ V}$$

3.5 ECL (EMITTER COUPLED LOGIC)

The ECL logic family is similar to TTL, but the transistors are arranged never to switch into the fully saturated on and off states. This increases switching speeds. However ECL is expensive and power dissipation is high. For these reasons ECL is only applied in specialised applications.

3.6 MOS (METAL OXIDE SEMICONDUCTOR)

MOS circuits offer an enormous leap in packing density over TTL—typically 1000 more circuits can be built into a silicon wafer. MOS is not offered as a competitor to TTL, and there is no MOS series of chips performing gating, counting and other functions as with the 7400 TTL series. Instead MOS is applied to fabricate the principal parts of microcomputer circuits within single chips, e.g. the microprocessor and memory chips. Each chip may contain tens of thousands of transistors.

ICM-C

Silicon dioxide insulator

Source

Gate

Metallic connections

Drain

n+   n channel   n+

p

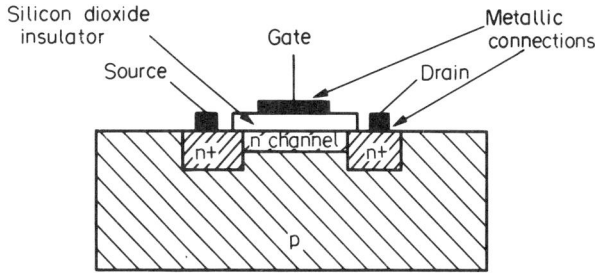

(a) Construction

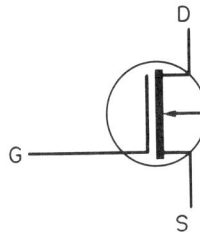

D

G

S

(b) Circuit symbol for depletion mode

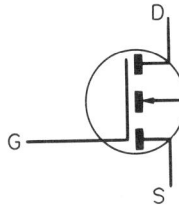

D

G

S

(c) Circuit symbol for enhancement mode

FIG. 3.10. Insulated gate FET (IGFET).

The principal difference between TTL and MOS is that a new type of transistor is used within MOS to perform gating, switching, etc. This is the field effect transistor (FET). TTL circuits are often termed "bipolar" circuits because they use the conventional transistor which passes current using two carriers—holes (positive electronic charge) and electrons (negative electronic charge). MOS is often termed "unipolar" because the FET employs only one carrier type (p or n).

Figure 3.10 shows an n-channel insulated gate FET (sometimes called IGFET). The principle of operation is that the channel conductivity between source and drain is varied by applying a voltage to the gate connection. The

$+ V_{DD}$

$X = \overline{A + B}$

A

$0\ V$

B

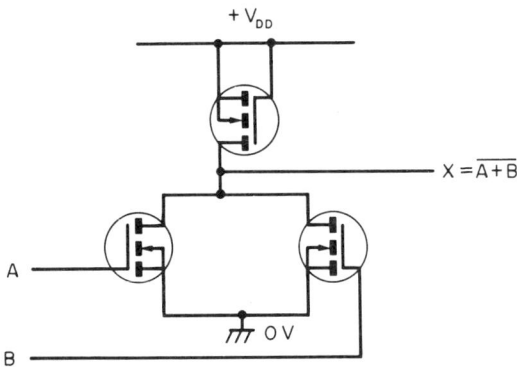

FIG. 3.11. MOS NOR gate.

device possesses an extremely high input impedance since the gate carries no current and the channel conductivity is affected by an electric field across the channel. The circuit symbol in (b) is for a device that operates in the "depletion" mode, i.e. increasing the reverse bias on the gate causes the channel to be restricted, or depleted. The symbol in (c) is for a device which operates in the "enhancement" mode, i.e. increasing the forward bias increases, or enhances, channel conductivity. The name MOS (Metal Oxide Semiconductor) is derived from the metallic/silicon dioxide insulating barrier at the gate connection.

MOS chips offer complex circuitry using tens of thousands (or even more) of such transistors. A microprocessor chip supports gates, counters, registers, storage elements, etc., and for example purposes only here, a simple MOS NOR gate is shown in Fig. 3.11. Notice that there are no resistors in the circuit—this leads to the much greater circuit packing density compared with TTL. The top FET in the circuit acts as a load resistor to the lower two FETs, which simply act as parallel switches—see switch circuit logic in section 3.2. When both A and B are at 0 V (logic 0), the lower FETs are both switched off and the output goes to $+V$ (logic 1). If either (or both) input goes high (logic 1) then one (or both) FETs switch on, and the output goes low (logic 0).

There are two classifications of MOS—PMOS (p channel) and NMOS (n channel). NMOS microprocessors have superseded the early PMOS devices due to a factor of ten increase in speed. Memory and input/output chips are also constructed using NMOS.

Normally MOS chips are TTL-compatible, i.e. they operate using power supply voltage levels of $+5$ V and 0 V and they can be interfaced directly to TTL devices.

3.7 CMOS (COMPLEMENTARY MOS)

CMOS is a development of MOS and employs both p and n type FETs connected in a "complementary" manner. CMOS possesses a significant advantage over MOS—a much lower (factor of greater than 1000) power consumption. However it is more expensive than MOS.

CMOS is not as popular as TTL and MOS, but it is offered as replacements for:

(a) TTL SN7400 series of gates, registers, etc.
(b) MOS microprocessors and memory chips.

The CMOS series of devices which is equivalent to the TTL SN7400 range is the "4000B" series, e.g. the 4011B (see Fig. 3.12) is a Quadruple 2-input NAND gate and is compatible with the TTL SN7400 (see section 2.4). The CMOS devices are used when lower power, improved noise immunity and a wider supply voltage range are required. Typically CMOS devices operate for a supply voltage range of from 3 to 15 V; TTL has an extremely narrow operating range (4.75 to 5.25 V). TTL and CMOS prices are comparable.

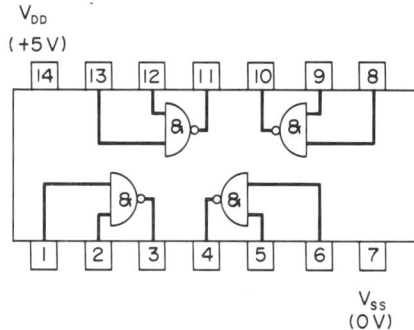

FIG. 3.12. CMOS Quadruple 2-input NAND gate (4011B).

A CMOS NOR gate is illustrated in Fig. 3.13. Notice that both n channel (lower FETs) and p channel (upper FETs) are used. When both A and B are at 0 V, the lower FETs are both switched off and the upper FETs are both switched on—this produces a logic 1 at the output. If either (or both) input goes high, then one (or both) of the lower FETs switches off—this produces a logic 0 at the output.

Noise immunity with CMOS circuits is good because logic levels are very close to the power supply levels (+5 V and 0 V normally).

CMOS circuits possess one disadvantage compared with TTL and MOS—

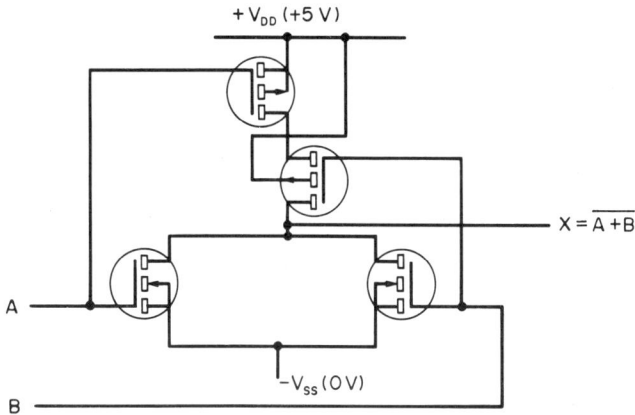

FIG. 3.13. CMOS NOR gate.

they must be handled with care to prevent damage by static electricity. Although most CMOS ICs possess internal protection against high static voltages, it is sensible to store them with their interconnecting pins embedded in conducting foam. Material and clothing of nylon and man-made fibres should be avoided, and all test equipment (often even including the operator of the test equipment) should be earthed. Unused inputs must not be left floating. They should be connected (typically through a 100 K resistor) to one of the supply lines.

Interfacing between CMOS and TTL devices sometimes demands special measures as follows:

(a) Connecting a TTL output to a CMOS input—no special measures need to be taken if the CMOS device operates with supply voltage levels of +5 V and 0 V, but a buffer, e.g. a 7407 open collector buffer-driver, is necessary if the dc supply voltage levels are different (perhaps +10 V and 0 V).

(b) Connecting a CMOS output to a TTL input—series A CMOS (4000A) cannot supply sufficient source and sink current to drive normal TTL, but series B CMOS (4000B) can drive low power TTL (74L00 and 74LS00) directly, CMOS buffers, e.g. 4010A, are available for normal TTL and these can change voltage levels as well as provide sufficient current drive.

High packing density CMOS chips which are designed to replace MOS devices, e.g. microprocessors, possess the advantage of lower power consumption. Applications are for battery-driven systems, e.g. digital watch (1.5 V battery), pocket calculator.

3.8 COMPARISON OF LOGIC FAMILIES

The principal features of the different IC logic families are summarised in Table 3.1.

TABLE 3.1   COMPARISON OF LOGIC FAMILIES

Feature	TTL	LS TTL	ECL	MOS	CMOS
1. Speed (propagation delay through gate)	10 ns	10 ns	2.5 ns	50 ns	35 ns
2. Power dissipation (per gate)	10 mW	2 mW	25 mW	0.1 mW	10 mW
3. Noise immunity	1 V	1 V	0.2 V	1 V	2 V
4. Fan-out	10	10	30	50	50
5. Size (degree of circuit integration)	MSI	MSI	MSI	VLSI	VLSI

Observe the following points:

1. The relatively slow switching speeds of MOS and CMOS prevent their use in the principal processing components of fast computers, e.g. multi-user mainframe computers.
2. CMOS has by far the lowest power consumption, making it the most suitable for battery-driven applications.
3. Noise immunity is the magnitude of an error noise signal which can be tolerated on the input signal to a circuit before incorrect circuit operation occurs. Again CMOS exhibits the best performance (although noise immunity is rarely a problem with TTL). Additionally CMOS operates over the widest supply range.
4. The fan-out of TTL is only 10, i.e. a TTL gate can drive up to 10 succeeding gates. MOS and CMOS possess far higher fan-outs due to their low current signals, e.g. CMOS signal currents are typically 10 pA (10 picoamps). However speeds are reduced for increased fan-outs.
5. The circuit integration, or packing density, abbreviations used in the table are defined as follows:

SSI—Small Scale Integration (less than 10 gates per chip),
MSI—Medium Scale Integration (between 10 and 100 gates per chip),
LSI—Large Scale Integration (between 100 and 1000 gates per chip),
VLSI—Very Large Scale Integration (greater than 1000 gates per chip).

One other circuit family which is not widely applied is I^2L (Injection Injection Logic), but it offers the following features:

(a) bipolar transistors,
(b) 8 ns propagation delay,
(c) VLSI.

It is applied with some memory devices, and will undoubtedly achieve more widespread application in the future.

BIBLIOGRAPHY

1. *Digital Techniques and Systems*, D. C. Green, Pitman, 1980.
2. *Microelectronics: a practical introduction*, R. A. Sparkes, Hutchinson, 1984.
3. *Digital Integrated Circuits and Computers*, Barry Woollard, McGraw-Hill, 1978.

EXERCISES

1. State three reasons why switch logic is not as suitable as electronic logic for the design of digital computers.
2. Examine the circuit of a TTL gate (Fig. 3.6) and follow the gating and switching action through each of the transistors for each combination of input logic levels.
3. Sketch the totem pole output stage of a TTL gate. What special external circuit arrangement is necessary if the gate output is "open collector"?
4. How many succeeding gates can a TTL gate drive normally?
5. What are the two significant advantages of LS series TTL devices over the standard TTL devices (see Appendix A)?
6. What are the standard TTL voltage logic levels?
7. What special arrangement should be made for unused inputs to a TTL gate, and why is this necessary?
8. What is the principal reason why a MOS gate can be constructed in a smaller circuit area in a silicon wafer than a TTL gate?
9. What is the name given to the series of CMOS devices that offer compatible functions to the TTL SN7400 series?
10. Why is it unwise for circuit assembly workers to wear nylon overalls when constructing circuits containing CMOS devices?
11. Would you choose TTL or CMOS circuits for a logic switching system in a satellite? Give a reason for your answer.
12. Would you choose TTL, ECL, MOS or CMOS for the central computing circuits of a multi-user computer, which must process programs as quickly as possible? Give a reason for your answer.
13. State four advantages of CMOS gating circuits over TTL.

14. Gain access to a circuit board within a home computer or office computer and identify the circuit family of each DIL. Use Appendix A to identify the circuit function of each TTL chip. Locate the dc power supply circuit (beware of mains voltage) and examine if it conforms to the common arrangement shown in Fig. 14.1.

CHAPTER 4

Bistables (Flip-flops) and Their Applications

(A bistable multivibrator, or "flip-flop", has two stable states—logic 1 and logic 0. The device can remain in either state for an indefinite period, until triggered into the alternative state by an input signal.)

4.1 S–R FLIP-FLOP

An S–R (Set–Reset) flip-flop is shown in Fig. 4.1. When both Set and Reset inputs are at logic 0, the flip-flop retains the previously stored bit (1 or 0), and the outputs stay unaltered. When Set is taken to logic 1 (and then released), the device stores a 1 and the output Q goes to 1 also. This state persists until Reset is set to 1, when the device clears to 0.

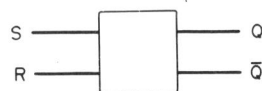

S	R	Q	$\overline{Q}$
O	O	No change	
I	O	I	O
O	I	O	I
I	I	Indeterminate	

(a) Circuit symbol   (b) Truth table

FIG. 4.1. S–R flip-flop.

Two methods of constructing this flip-flop are shown in Fig. 4.2. The reader may like to test his knowledge of the simple NOR and NAND gating functions by verifying the truth table—it is suggested that you indicate the logic levels around the circuit for each of the middle two lines of the truth table.

(a) Using NOR gates

Notice that Q and Q̄
are reversed
cf. Fig. 4.1

(b) Using NAND gates

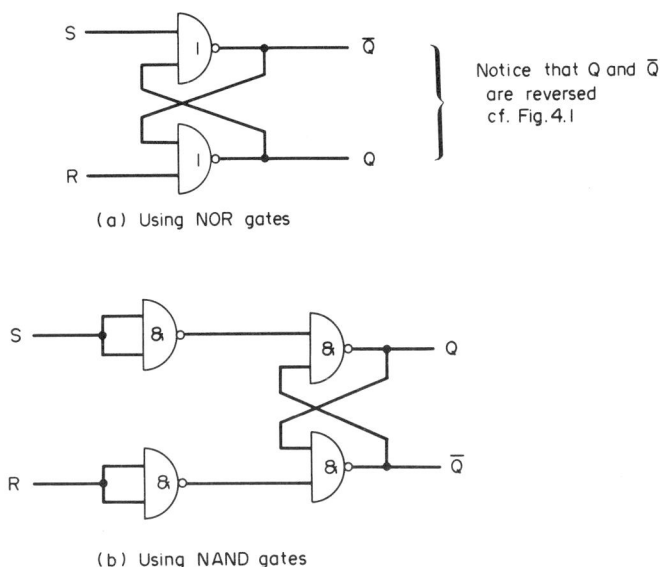

FIG. 4.2. S–R flip-flop constructed using NOR and NAND gates.

4.2 CLOCKED S–R FLIP-FLOP

In this case a third input (clock) is applied, and no change occurs within the flip-flop until this input is at logic 1. A clocked S–R flip-flop can be constructed using NAND gates as shown in Fig. 4.3. The truth table in Fig. 4.1 only applies when this CLK input changes to 1, because the outputs of the first two NAND gates only correspond with the S and R inputs when CLK is at 1.

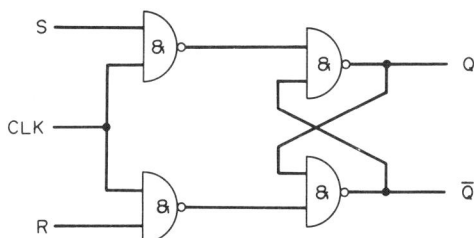

FIG. 4.3. Clocked S–R flip-flop constructed using NAND gates.

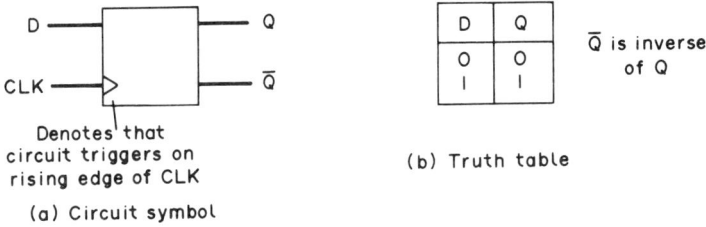

D	Q
0	0
1	1

$\overline{Q}$ is inverse of Q

Denotes that circuit triggers on rising edge of CLK

(b) Truth table

(a) Circuit symbol

Fig. 4.4. D-type flip-flop.

4.3 CLOCKED D-TYPE FLIP-FLOP

The D-type flip-flop overcomes the problem of the indeterminate state of the S–R type when both of its inputs are at 1; it achieves this by ensuring internally that S and R are always the complement of each other. Additionally the D-type flip-flop is offered in a single IC package, and it does not have to be constructed using NOR or NAND gates.

Figure 4.4 shows the circuit symbol and truth table for the D-type flip-flop. The output Q only changes to the states shown when the CLK (clock) input changes to 1. Notice how this rising edge (changing to 1) triggering state is indicated on the diagram.

The D-type flip-flop is applied commonly in storage registers and shift registers, as described later in this chapter. One simple application is to divide the frequency of a pulse stream by two, as shown in Fig. 4.5. The output can be taken either from Q or $\overline{Q}$.

A practical example of a D-type flip-flop IC is the 7474 (or SN7474), which

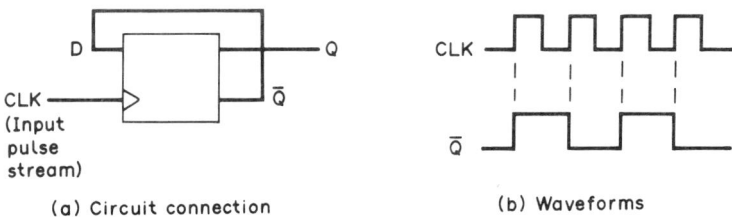

CLK (Input pulse stream)

(a) Circuit connection

(b) Waveforms

Fig. 4.5. Divide by two using D-type flip-flop.

V$_{cc}$

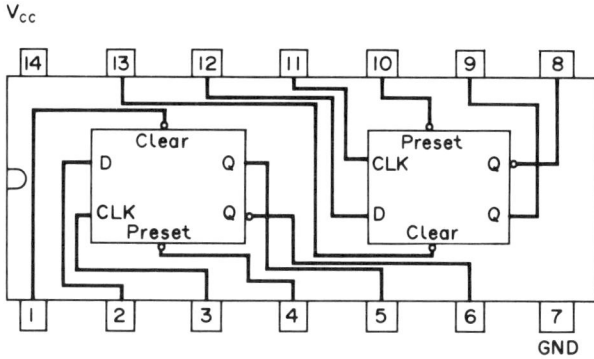

FIG. 4.6. Dual D-type flip-flop (7474).

supports two flip-flops and is illustrated in Fig. 4.6. Notice that there are additional "over-ride" inputs—CLEAR and PRESET. The former sets the flip-flop to the 0 state ($Q = 0$), whilst the latter sets the device to the 1 state ($Q = 1$). The inverse logic nature (set when at logic 0) of some signals in this diagram is illustrated by the bubble at their connection point.

4.4 CLOCKED J–K FLIP-FLOP

The clocked J–K flip-flop is shown in Fig. 4.7. The output Q only changes to the states shown when the CLK input changes to 1, i.e. on the rising edge of CLK. The word "toggles" means that the output reverses, e.g. 1 to 0 or 0 to 1. The J–K flip-flop is used frequently in this mode, i.e. with the J and K inputs held permanently to 1, e.g. in counters. A single J–K flip-flop can be used to divide a pulse stream by two, as shown in Fig. 4.8.

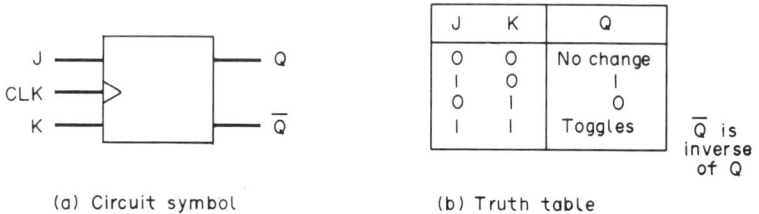

J	K	Q
0	0	No change
1	0	1
0	1	0
1	1	Toggles

$\overline{Q}$ is inverse of Q

(a) Circuit symbol   (b) Truth table

FIG. 4.7. J–K flip-flop.

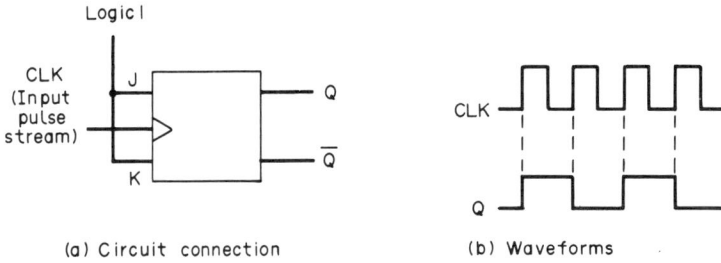

(a) Circuit connection   (b) Waveforms

FIG. 4.8. Divide by two using J–K flip-flop.

A practical example of a J–K flip-flop IC is the 7473 (or SN7473), which supports two flip-flops and is illustrated in Fig. 4.9. Notice that each flip-flop possesses a CLEAR function but no PRESET. If it is required to reverse these functions so that the device has a PRESET but no CLEAR function, then this is achieved by reassigning J as K and K as J; Q and $\overline{Q}$ are also interchanged. The bubble at the CLK input in the diagram indicates that the flip-flop is negative-edge triggered, i.e. it triggers on the falling edge of CLK.

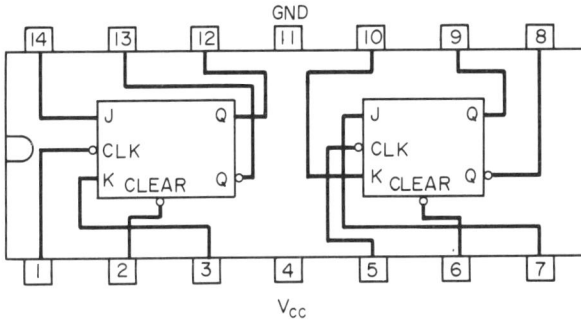

FIG. 4.9. Dual J–K flip-flop (7473).

In order to avoid a proliferation of flip-flop types in a system, it is sometimes convenient to use a J–K flip-flop connected as a D-type—perhaps one-half of a dual J–K IC is spare in a circuit application. This is performed as shown in Fig. 4.10.

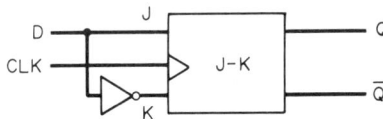

FIG. 4.10. J–K flip-flop connected as a D-type flip-flop.

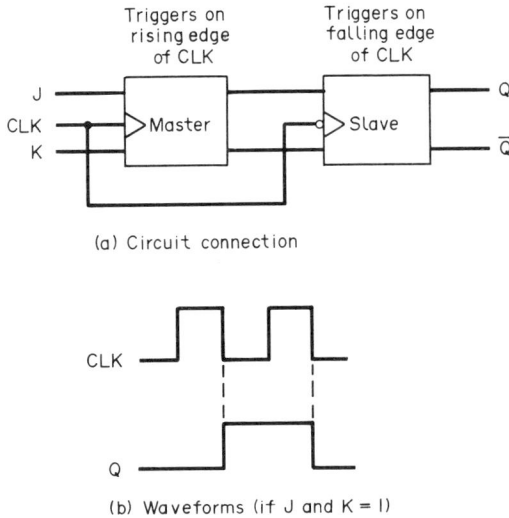

(a) Circuit connection

(b) Waveforms (if J and K = 1)

FIG. 4.11. Master-slave J–K flip-flop.

4.5 MASTER–SLAVE FLIP-FLOP

The clocked flip-flops examined so far (clocked S–R, clocked D-type and clocked J–K) are all edge-triggered. This can cause problems when flip-flops are cascaded together in a counter with feedback applied, and incorrect triggering of individual flip-flops can occur. This problem is solved by the use of master-slave flip-flops, which require a full pulse (leading edge *and* trailing edge) on the CLK input before they trigger. A master-slave J–K flip-flop, which consists of two J–K flip-flops connected in series, is illustrated in Fig. 4.11. The first (master) flip-flop triggers on the rising edge of the CLK input, and the second (slave) flip-flop triggers on the falling edge of CLK. Therefore the entire flip-flop requires a full pulse on the CLK input before it triggers.

An example of a practical master-slave flip-flop IC is the 7472, which supports only a single flip-flop circuit.

4.6 STORAGE REGISTER

A storage register is a circuit that stores a group of bits and holds it for later use. D-type flip-flops are normally used to construct registers, and a simple

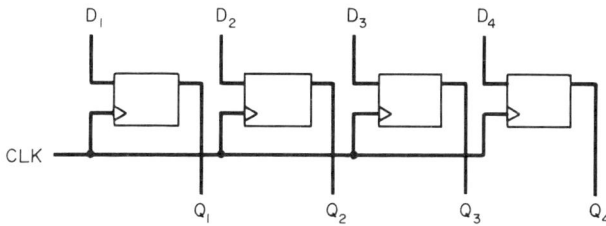

FIG. 4.12. 4-bit storage register using D-type flip-flops.

4-bit register is shown in Fig. 4.12. The CLK signal gates the four input bits D_1 to D_4 into the register, which presents these stored bits on the outputs Q_1 to Q_4. The input signals can be cleared or altered, but this will not affect the output bits until another CLK pulse occurs.

A very useful practical storage register IC is the 74373 (SN74373), which is illustrated in Fig. 4.13. This chip is used frequently in microcomputers as an "output port", when it staticises 8 bits which are connected to an external device or circuit, e.g. a segment display, or 8 LEDs. The 8 outputs pass

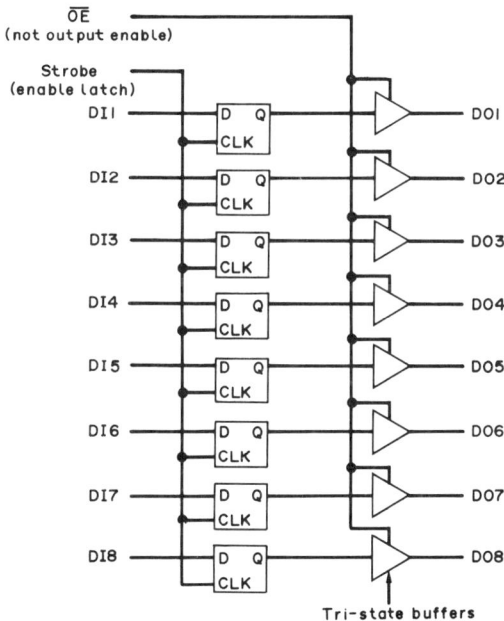

FIG. 4.13. 8-bit storage register (74373).

through "*tri*-state" buffers within the chip. In the normal configuration the $\overline{OE}$ pin, which enables these buffers, is held at logic 0, i.e. the outputs are permanently enabled and the Q outputs from the flip-flops pass through to the output pins. If $\overline{OE}$ should be set to logic 1, the output pins are in the "floating" state, i.e. they are electrically disconnected from the flip-flop Q outputs and they carry neither 1s or 0s. The STROBE input connects to all the CLK connections on the flip-flops, and gates in the input bits to the register. An alternative name for this IC is "octal latch".

4.7 SHIFT REGISTER

A shift register is a register which is fed with data bits in serial form and outputs those bits in parallel form, or vice versa. A simple 4-bit shift register is shown in Fig. 4.14. If logic 1 is applied at IN at the same time as a CLK pulse occurs, then that bit is set into the first flip-flop. That logic 1 passes from one flip-flop to the next on succeeding CLK pulses. If the input changes between 0 and 1 for four consecutive CLK pulses, then that 4-bit input pattern is shifted into the register and appears in parallel on the outputs Q_1 to Q_4. Therefore the circuit operates as a serial to parallel converter. A parallel to serial converter circuit is very similar to this arrangement, but allows the parallel data to initialise the flip-flops using the PRESET and CLEAR signals. This bit pattern is then pulsed out of the cascaded line of flip-flops in serial form.

A shift register can operate in one of the following four ways:

(a) serial in, parallel out,
(b) parallel in, serial out,
(c) serial in, serial out,
(d) parallel in, parallel out.

Shift registers, which are not as commonly applied as simple storage registers, can be constructed from several discrete flip-flops. Alternatively they are available as single ICs in the standard TTL range, e.g. the 74194 is a 4-bit bi-directional shift register, whilst the 74164 and 74165 are 8-bit serial-to-parallel and parallel-to-serial shift registers respectively.

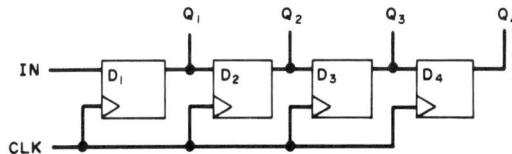

Fig. 4.14. 4-bit shift register.

4.8 ASYNCHRONOUS COUNTERS (RIPPLE-THROUGH COUNTERS)

An asynchronous counter is a tandem connection of flip-flops (normally J–K). The first pulse in the incoming pulse stream which is to be counted initiates the switching of the first flip-flop, but switching of each of the other flip-flops is initiated by the preceding flip-flop. For this reason the system is often called a "ripple-through counter", as the carry bit from each stage "ripples" through the counter.

Figure 4.15 illustrates a 4-bit ripple-through counter. Each master-slave flip-flop is toggled on the falling (or negative) edge of the pulse applied to its CLK input. It can be seen that the Q outputs produce a binary count of the number of input pulses, as shown in Table 4.1.

The outputs can be used to drive indicator LEDs or a segment numerical display.

This 4-bit counter resets itself to 0 after 16 pulses. However it can be converted such that it counts to any chosen value before reset occurs. Clearly the most common application is as a decade counter, such that it resets itself

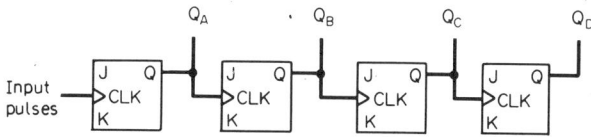

Note All J and K inputs are connected permanently to I

(a) Circuit connection

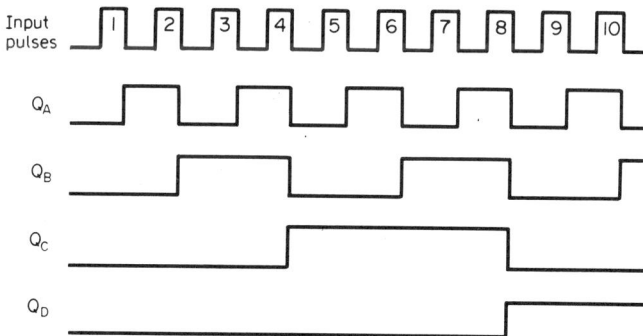

Note Flip-flops are master-slave, and so trigger after full input pulse

(b) Waveforms

FIG. 4.15. Asynchronous (ripple-through) counter.

ICM-D

TABLE 4.1   RIPPLE-THROUGH COUNTER
OUTPUTS

Input Pulses	Outputs			
	Q_D	Q_C	Q_B	Q_A
0	0	0	0	0
1	0	0	0	1
2	0	0	1	0
3	0	0	1	1
4	0	1	0	0
5	0	1	0	1
6	0	1	1	0
7	0	1	1	1
8	1	0	0	0
9	1	0	0	1
10	1	0	1	0

after 10 pulses, as shown in Fig. 4.16. The Q_B output is set after every 2 input pulses and the Q_D output is set after every 8 input pulses. They are set together therefore after 10 pulses, as confirmed in Table 4.1, and the output of the AND gate is used to clear the whole counter when a count of 10 is reached.

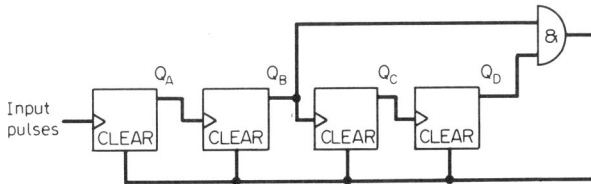

Fig. 4.16.   4-bit counter connected as a decade counter.

Decade counters can be connected in tandem if it is required to count a larger number of pulses, as illustrated in Fig. 4.17. The maximum number of

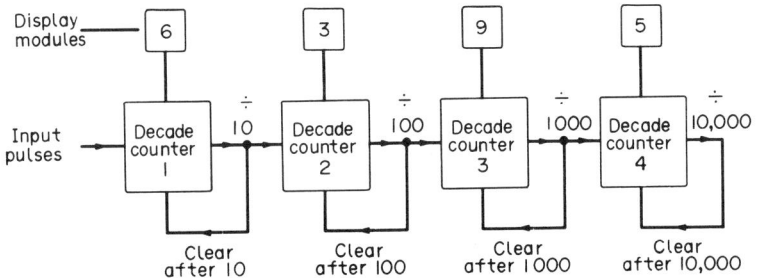

Fig. 4.17.   4-decade counter.

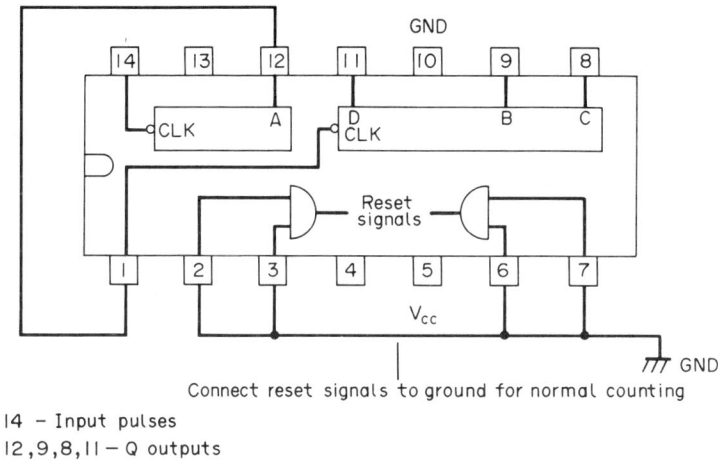

FIG. 4.18. Decade counter (7490).

I4 – Input pulses
I2,9,8,II – Q outputs

pulses that can be counted is 9999 in this arrangement, and the current count attained is shown as 5936.

4-bit counters are available as single ICs in the TTL 7400 series, e.g.

(a) 7493, which is a simple 4-bit binary counter—this chip is illustrated in a full counter/display system in section 5.5.
(b) 7490, which is a decade counter—its pin functions are shown in Fig. 4.18 (note that the connection from the first stage to the other three stages must be made externally, and also that four external resets can be used to make the device reset after counts other than 10).

4.9 SYNCHRONOUS COUNTERS

There is an inherent small switching, or propagation, delay in a ripple-through counter, because the preceding stages must switch before a later flip-flop switches. This causes problems if the input pulses are very fast, and for this reason the synchronous counter has been developed. It is more complex and more expensive than the asynchronous (ripple-through) counter.

Figure 4.19 shows the circuit arrangement. In this configuration all the flip-flops which are required to change state do so simultaneously; there is no ripple-through effect. Only the J and K inputs to the first flip-flop are connected permanently to 1, and the other flip-flops have their J and K inputs set by the AND function of the outputs of all previous flip-flops. The

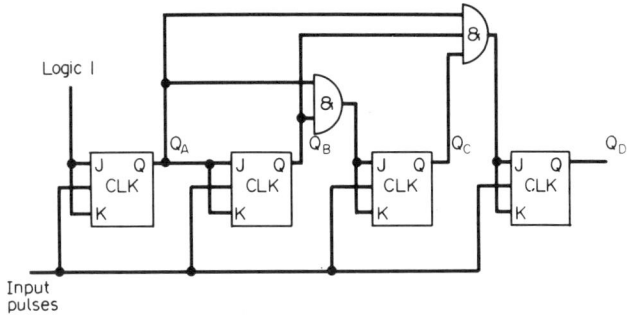

FIG. 4.19. Synchronous counter.

waveforms are exactly the same as shown for the ripple-through counter—see Fig. 4.15(b).

An example of one stage in the counting sequence is when the fourth input pulse occurs. The binary count after the third pulse is:

$$\begin{array}{cccc} Q_A & Q_B & Q_C & Q_D \\ 1 & 1 & 0 & 0 \end{array}$$

and so the fourth pulse causes the Q_C flip-flop to toggle to the 1 state because its J and K inputs are at 1. Similarly the seventh pulse leaves:

$$\begin{array}{cccc} Q_A & Q_B & Q_C & Q_D \\ 1 & 1 & 1 & 0 \end{array}$$

and so the eighth pulse causes the final Q_D flip-flop to toggle to 1.

A practical synchronous counter IC is the TTL 74193, which is a binary synchronous up/down counter—see Fig. 4.20. A binary count can be preset

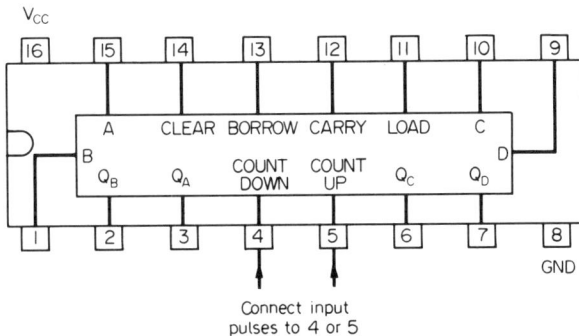

FIG. 4.20. Up/Down synchronous counter (74193).

into the counter on A, B, C and D if LOAD is held at logic 0. Input pulses must be connected either to pin 4 or pin 5 (the unused input must be set high) and the outputs are on Q_A, Q_B, Q_C and Q_D. BORROW and CARRY can be connected to COUNT DOWN and COUNT UP inputs of a subsequent counter if included.

BIBLIOGRAPHY

1. *Digital Design*, M. Morris Mano, Prentice/Hall, 1984
2. *Digital Electronics*, Robert E. Genn, Prentice/Hall, 1982.
3. *Digital Electronic Technology TEC Level IV*, D. C. Green, Pitman, 1982.

EXERCISES

1. What is the "indeterminate" state in a S–R flip-flop?
2. Sketch the missing waveform at the output connection Q for each of the following flip-flops:

   (a) *S–R*

   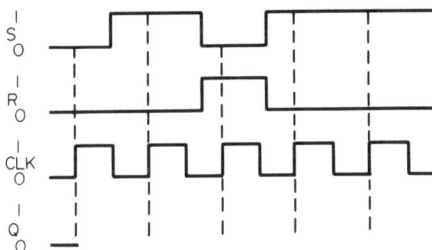

   (b) *D-type*

   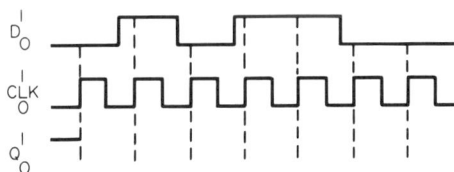

   (c) *J–K*

   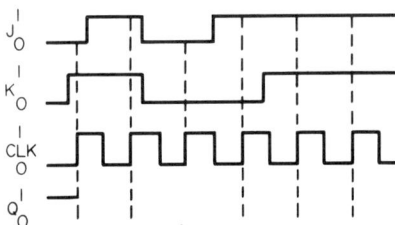

3. What two "over-ride" signals are often used with flip-flops?
4. What is the difference in triggering requirements for a master-slave flip-flop compared with other flip-flops?
5. Sketch three circuits (using a D-type, J–K and master-slave J–K flip-flop) to divide an input pulse stream by 2.
6. Examine Fig. 4.14 and re-draw the circuit arrangement, showing the use of the PRESET and CLEAR signals, to produce a parallel in, serial out, shift register.
7. Re-draw Fig. 4.16 such that the counter resets after 5 input pulses.
8. State an advantage and a disadvantage of a synchronous counter compared with an asynchronous counter.
9. How can a 3-stage decade counter be used to light an indicating lamp when 199 input pulses have occurred? Sketch the circuit, which should employ a triple-input AND gate.
10. Examine the pin layout of the SN74669 in Appendix A and describe the function of each interconnection signal.
11. What are the outputs of the following circuit after 36 input pulses?

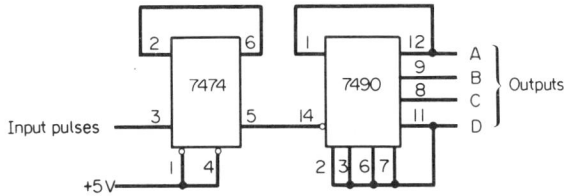

CHAPTER 5

Logic Systems

5.1 ASTABLE MULTIVIBRATOR

An astable multivibrator is a circuit that has no stable states and continually switches between two unstable states. It is used as a pulse generator.

A simple astable multivibrator can be constructed using two inverters as shown in Fig. 5.1. The output of each inverter is fed back to the input of the other inverter. The role of the capacitor is to delay the effect of the voltage output from one inverter before it switches the other inverter. The inverters repeatedly switch 1 and 0 outputs. The durations of the on and off times ("mark" and "space") for the pulses produced are determined by the R and C values. Circuit (a) gives full control of both times; circuit (b) gives control of only one time. If potentiometers (variable resistors) are used in place of resistors, then a variable pulse duration and frequency can be achieved.

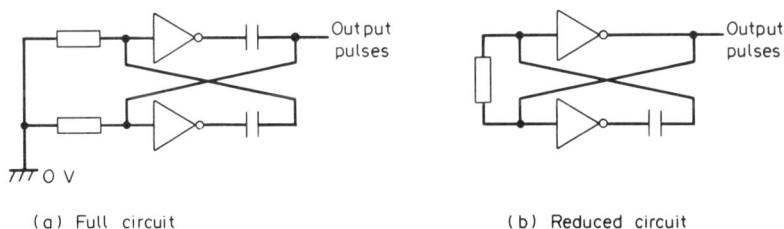

(a) Full circuit   (b) Reduced circuit

FIG. 5.1. Astable multivibrator using inverters.

Sometimes it is required to generate pulses of a precise frequency, and the quartz crystal controlled multivibrator in Fig. 5.2 can be applied. The crystal locks the frequency to a chosen value. This type of circuit is used to generate the clock signal in a microcomputer.

One of the most popular and flexible ICs that is applied as an astable multivibrator is the 555 timer. The external components that must be

45

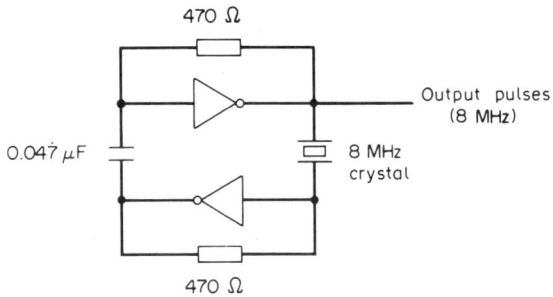

FIG. 5.2. Crystal-controlled astable multivibrator.

connected to this chip to cause it to operate in the "free-running" (astable) mode are shown in Fig. 5.3. The choice of these component values determines the mark and space times for the pulse sequence produced, as follows:

$$T_1 = 0.7(R_A + R_B)C$$
$$T_2 = 0.7R_BC$$

The capacitor which is connected to pin 5 eliminates noise.

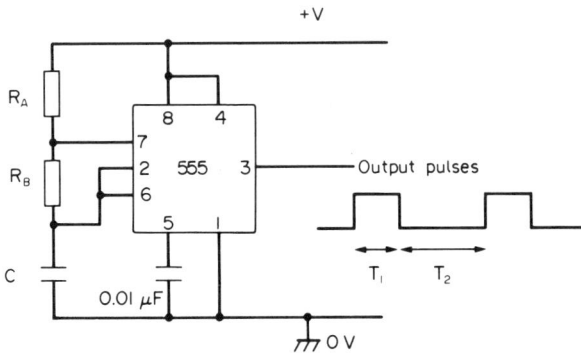

FIG. 5.3. Astable multivibrator using 555 timer.

5.2 MONOSTABLE MULTIVIBRATOR

A monostable multivibrator (or "one-shot") circuit has only one stable state, and it remains in the unstable state for a fixed time. It is used to generate a single fixed length pulse when triggered by an input pulse.

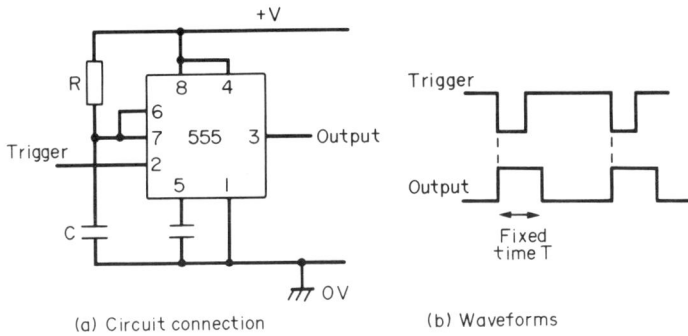

(a) Circuit connection   (b) Waveforms

FIG. 5.4. Monostable multivibrator using 555 timer.

Figure 5.4 shows the circuit connections required to produce a monostable multivibrator using the 555 timer chip. The circuit triggers on the negative-going input signal, and the time that the output stays in the high state is:

$$T = 1.1 \, RC$$

For example, if $R = 1 \, M\Omega$ and $C = 1 \, \mu F$, then the output pulse width is just over 1 second. The one-shot is not used as frequently as the astable multivibrator, but it can be applied to regenerate a fixed length pulse at the end of a transmission line which exhibits noise and signal loss.

5.3 SCHMITT TRIGGER

A Schmitt trigger circuit is used to convert a signal of irregular waveshape into a sharply defined pulse. Figure 5.5 shows the circuit symbol for a Schmitt

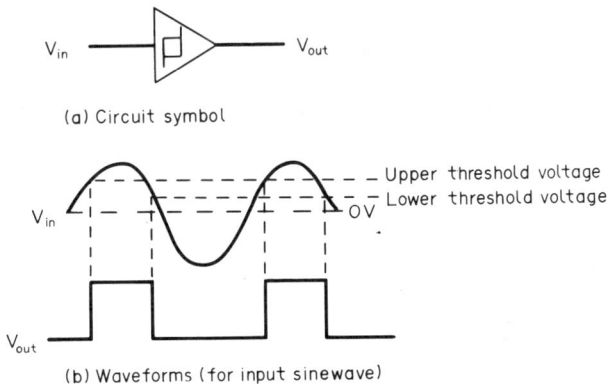

(a) Circuit symbol

(b) Waveforms (for input sinewave)

FIG. 5.5. Schmitt trigger.

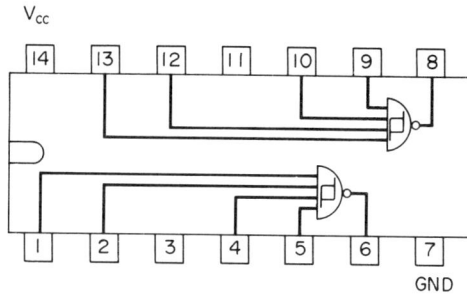

FIG. 5.6. Dual 4-input NAND gate with Schmitt trigger input (7413).

trigger, and the effect of the circuit on an input voltage sinewave signal. The circuit triggers on at the higher threshold voltage and off at the lower threshold voltage. Therefore it converts the sinewave to a squarewave. Other applications are to "clean up" noisy pulses.

A practical Schmitt trigger circuit is the TTL 7414. This offers six inverting Schmitt triggers in a 14-pin package.

A Schmitt trigger input circuit is offered with a NAND gate in the 7413, which is illustrated in Fig. 5.6. Each NAND gate can be triggered by slow-ramping or noisy signals to produce a jitter-free output signal. The "hysteresis", or "backlash", is the difference between the higher and lower threshold voltages, and is typically 0.8 V for this circuit.

An alternative circuit to the straightforward digital Schmitt trigger is a circuit based on an op-amp; this is described in section 8.7.

5.4 DECODERS

A decoder circuit converts a coded input bit pattern into a unique output bit pattern. Decoder ICs are used for the following purposes:

(a) binary input, decimal output (e.g. 2-bit input, 4 discrete output lines with only one output line set at any time),

(b) binary input, 7-segment pattern output (for numerical display).

The operation of a 2 to 4 decoder is illustrated in Fig. 5.7. Only one of the

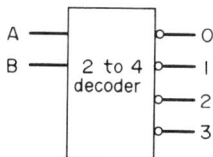

(a) Circuit symbol

Inputs		Outputs			
A	B	0	1	2	3
0	0	0	1	1	1
1	0	1	0	1	1
0	1	1	1	0	1
1	1	1	1	1	0

(b) Truth table

FIG. 5.7. 2 to 4 decoder.

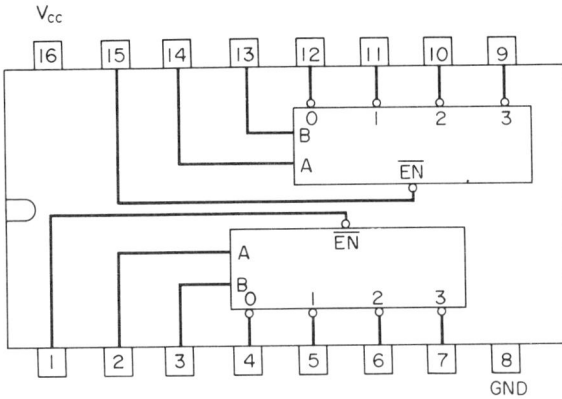

FIG. 5.8. Dual 1 of 4 decoder (74139).

four outputs is set at any time (set to logic 0 because the outputs are inverse logic signals)—the particular output line which is set is determined by the binary code on the two inputs. Notice the increasing binary count on the two inputs (A is the least significant bit). This device has application in micro-computer circuits when it is required to select only one of four memory chips (or input/output chips). Each of these chips is fed with one of the decoder output signals, as described in chapter 11.

A practical 2 to 4 decoder IC is the 74139, which is shown in Fig. 5.8. Two decoder circuits are supported on the same chip. Notice that the outputs are only set in the manner indicated in the truth table of Fig. 5.7(b) when the $\overline{EN}$ signal is set to logic 0.

Another popular decoder circuit is a 3 to 8 decoder, which is illustrated in Fig. 5.9. Again this device is widely applied in microcomputer circuits, when up to eight memory (or input/output) chips are applied but only one must be selected at any time. The increasing binary count on A, B and C is shown in

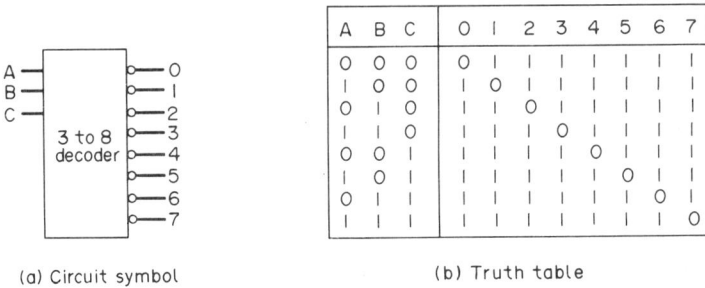

A	B	C	0	1	2	3	4	5	6	7
0	0	0	0	1	1	1	1	1	1	1
1	0	0	1	0	1	1	1	1	1	1
0	1	0	1	1	0	1	1	1	1	1
1	1	0	1	1	1	0	1	1	1	1
0	0	1	1	1	1	1	0	1	1	1
1	0	1	1	1	1	1	1	0	1	1
0	1	1	1	1	1	1	1	1	0	1
1	1	1	1	1	1	1	1	1	1	0

(a) Circuit symbol   (b) Truth table

FIG. 5.9. 3 to 8 decoder (1 of 8 decoder).

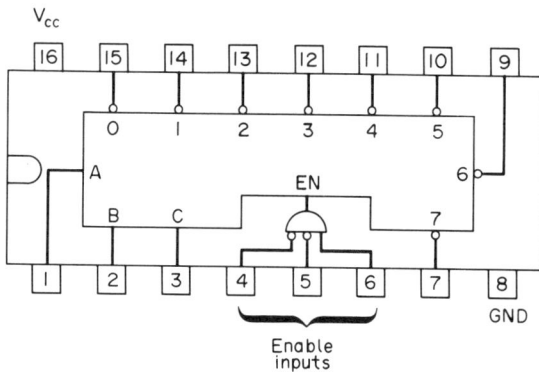

FIG. 5.10. 1 of 8 decoder (74138).

reverse order in the truth table. This circuit is available as the 74138, as shown in Fig. 5.10. The decoder only sets its outputs as indicated in the truth table when the Enable inputs are set as follows:

pin 4 – 0
pin 5 – 0
pin 6 – 1

Another type of decoder which is applied with segment displays is a binary to 7-segment decoder. Segment displays are used widely to display numerical values. If larger numbers of segments are used to construct characters then alphabet letters can be displayed with acceptable clarity. A 7-segment display is indicated in Fig. 5.11. The device can be constructed using LEDs, or it can be a liquid crystal display (LCD). The latter activates its segments using reflected light and is an extremely low-powered device. Both types of device are available as single digit display units or multiple digit units.

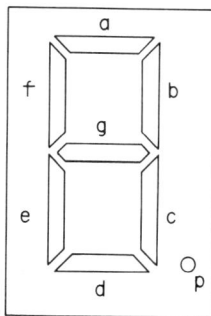

Note

(1) a to g are 7 segments
(2) p is decimal point
    (not normally used)

FIG. 5.11. 7-segment display.

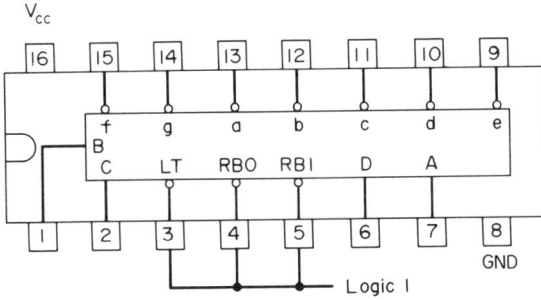

(a) Pin layout

Decimal	Inputs				Outputs							Segment
0	D	C	B	A	a	b	c	d	e	f	g	character
0	0	0	0	0	1	1	1	1	1	1	0	
1	0	0	0	1	0	1	1	0	0	0	0	
2	0	0	1	0	1	1	0	1	1	0	1	
3	0	0	1	1	1	1	1	1	0	0	1	
4	0	1	0	0	0	1	1	0	0	1	1	
5	0	1	0	1	1	0	1	1	0	1	1	
6	0	1	1	0	1	0	1	1	1	1	1	
7	0	1	1	1	1	1	1	0	0	0	0	
8	1	0	0	0	1	1	1	1	1	1	1	
9	1	0	0	1	1	1	1	1	0	1	1	

(b) Truth table

FIG. 5.12. Binary to 7-segment decoder (7447).

A binary to 7-segment decoder integrated circuit is the 7447, which is illustrated in Fig. 5.12. Normally pins 3, 4 and 5 are held at logic 1. Notice that the outputs are inverse logic, i.e. each segment bit is set when it is at logic 0. The truth table shows the signals set using normal positive logic for simplicity—these settings are inverted at the interconnection pins.

5.5 PULSE GENERATOR/COUNTER/DECODER/DISPLAY SYSTEM

At this point it may assist the reader to examine a large interconnected IC system that incorporates several of the devices described so far. Such a

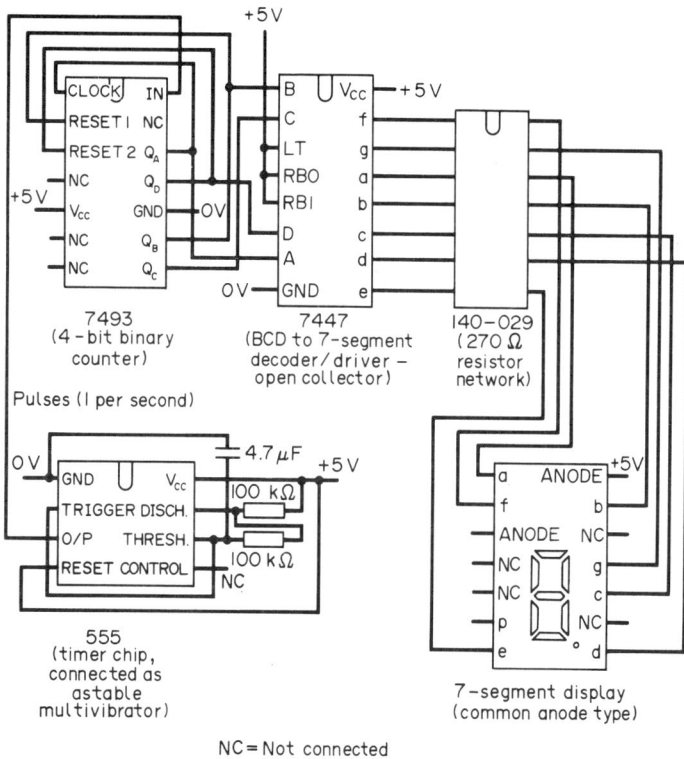

FIG. 5.13. Example circuit to count and display number of pulses.

system is illustrated in Fig. 5.13. The five IC stages in this example circuit are:

(a) Pulse generator, or astable multivibrator, using a 555 timer to generate one pulse every second (approximately)—see section 5.1.

(b) 4-bit binary counter, which is wired as a decade counter with the two RESET signals connected to the outputs Q_B and Q_D—see section 4.8 and Fig. 4.16.

(c) Binary (or BCD—Binary Coded Decimal) to 7-segment decoder, as described in section 5.4. BCD simply identifies the input binary number as a 4-bit number, which can therefore take decimal values 0 to 15.

(d) Resistor pack, which is a DIL package possessing a network of seven resistors and is applied in place of discrete resistors to reduce space on circuit boards.

(e) 7-segment display, which is a LED network mounted in a DIL package. All LED anodes are connected internally, so logic 0 applied to each cathode by the decoder chip draws current through that LED and its associated external 270 Ω resistor to illuminate the LED.

Notice that a resistor is required at each output of the decoder chip because its outputs are "open-collector" (see section 3.4 and Fig. 3.8). The resistor is connected through a LED in the following segment display unit to +5 V. The resistor limits the current through the LED to a value of approximately 20 mA, which produces average illumination in the LED.

This test circuit counts the generated pulses from 0 to 9, when the counter resets itself, and the count repeats. The RC component values in the pulse generator circuit (555 timer) are chosen to produce an extremely slow pulse rate so that the count can be observed on the display unit.

5.6 COMBINATIONAL LOGIC

The term combinational logic is applied to a gating system in which all signals at each point in the system change, or are likely to change, at the same time. It contrasts with a sequential logic system in which signals throughout the system change at different times; an example is an asynchronous counter (see section 4.8).

The most common application of combinational logic is in simple control circuits, for example:

(a) lift/elevator control,
(b) vending machine,
(c) industrial control of conveyors, hoists, general electrical machinery,
(d) burglar alarm system,
(e) traffic light control,

and many others.

Consider the logic gate system of Fig. 5.14. This circuit generates a 1

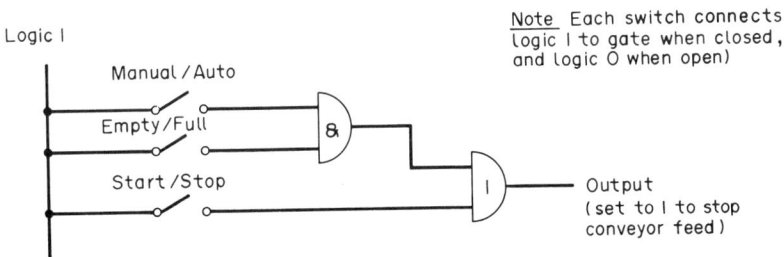

Fig. 5.14. Conveyor stop logic system.

output to stop a conveyor which feeds a bunker when:

either (1) a Manual/Auto switch is set to Auto

and

an Empty/Full limit switch is set to Full

or (2) a Start/Stop switch is set to Stop.

Sometimes it is required to standardise on a particular type of logic gate within a system, and Fig. 5.15 shows the same overall circuit function implemented using only NAND gates. The AND function is created using a NAND gate followed by an inverter (the latter is created using another NAND gate). The OR function is created using de Morgan's rule (see section 2.7), which states:

$$\overline{D + C} = \overline{D} . \overline{C}$$

and inverting both sides gives:

$$D + C = \overline{\overline{D} . \overline{C}}$$

Notice that this circuit can be simplified by the elimination of the inverters immediately before and after the signal D—double inversion restores the original logic level of course.

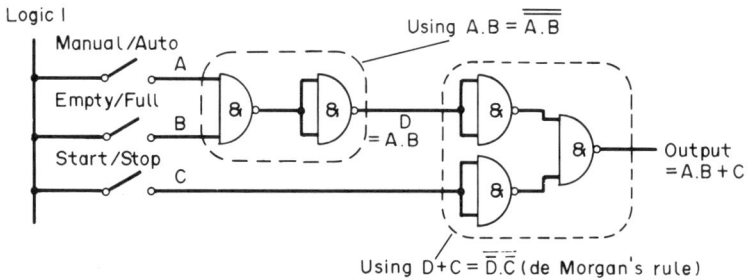

FIG. 5.15. Conveyor stop system implemented using only NAND gates.

Another example of a combinational logic system is shown in Fig. 5.16. this is a representation of part of the gating system required for a drinks dispensing machine, which dispenses a drink when:

(1) water is available,

and (2) paper cup is in position,

and (3) correct coin is inserted,

and (4) tea or coffee is selected.

A final example of a combinational logic system includes the use of a D-type flip-flop and is shown in Fig. 5.17. This circuit monitors the setting of

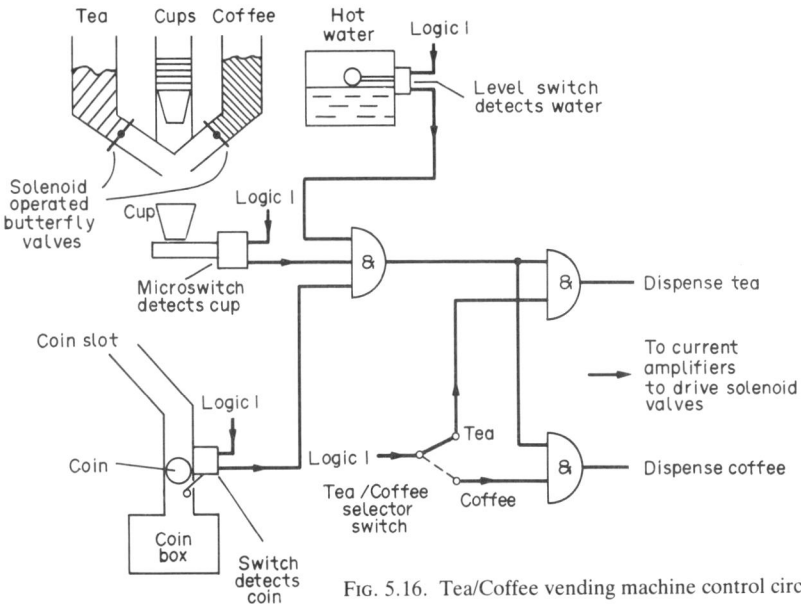

Tea   Cups Coffee   Hot   Logic I
                      water

Level switch
detects water

Solenoid
operated   Cup   Logic I
butterfly
valves

Microswitch
detects cup

&

Dispense tea

To current
amplifiers
to drive solenoid
valves

Coin slot

Logic I

Coin

Logic I   Tea

&   Dispense coffee

Tea/Coffee   Coffee
selector
switch

Coin
box   Switch
         detects
         coin

FIG. 5.16. Tea/Coffee vending machine control circuit.

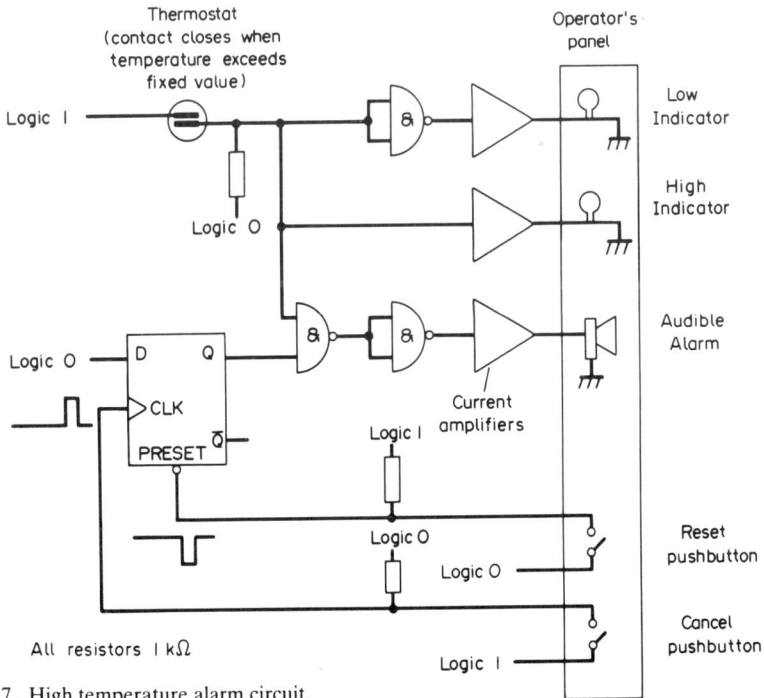

Thermostat   Operator's
(contact closes when   panel
temperature exceeds
fixed value)

Logic I   Low
                                               Indicator

                                               High
Logic O   Indicator

Logic O   D   Q   Audible
                                               Alarm

         CLK   Current
                 Logic I   amplifiers
         PRESET Q

                 Logic O   Reset
                                               pushbutton
                      Logic O

All resistors I kΩ   Cancel
                                               pushbutton
                 Logic I

FIG. 5.17. High temperature alarm circuit.

a thermostat (bi-metal strip), which changes from open-contact to closed-contact when the temperature exceeds a fixed value. The Low Indicator lamp is lit when the temperature is below this value, but when the temperature exceeds this value:

(a) the High Indicator lamp is lit,

and (b) the Audible Alarm is sounded.

The flip-flop is applied to enable the operator to cancel the Audible Alarm, but to leave the High Indicator lamp lit. If the flip-flop is initially set to the logic 1 state by pulsing the PRESET input to 0, the Q output is set to 1. This enables the Audible Alarm circuit. If the Cancel pushbutton is pressed, pulsing the CLK input to 1, when the Audible Alarm is sounding, the Q output is set to 0. This inhibits the Audible Alarm circuit.

Notice that this circuit shows that a resistor must be applied to pull the gate input to logic 0 for the thermostat circuit when the contact-closure signal is not set to closed. Clearly the logic 1 overrides the resistor connection to logic 0 when the contact is closed. This arrangement is necessary because it is assumed that TTL NAND gates, e.g. 7400, are used, and unconnected TTL inputs are set to logic 1 (see section 3.4).

BIBLIOGRAPHY

1. *Digital Integrated Circuits*, Joseph E. Kasper and Steven A. Feller, Prentice/Hall, 1983.
2. *Engineering Approach to Digital Design*, William I. Fletcher, Prentice/Hall, 1980.
3. *Digital Techniques*, Noel M. Morris, Macmillan, 1979.

EXERCISES

1. Design an astable multivibrator using a 555 timer which generates pulses at 330 μs duration at a repetition rate of 30 pulses/second. (Hint: try C = 0.047 μF)

2. Sketch a circuit using a 555 timer to generate a pulse of duration 1 ms every time an input signal changes from logic 1 to 0.

3. What is the definition of the hysteresis of a Schmitt trigger?

4. What is the state of the output of the following circuit when the input goes to logic 0?

5. What inputs must be set on pins 1 to 6 to produce a logic 0 output on pin 10 on a 74138 3 to 8 decoder?

6. What do you think would be the effect on the display if the common anode 7-segment display unit in Fig. 5.13 is replaced with a common cathode type?

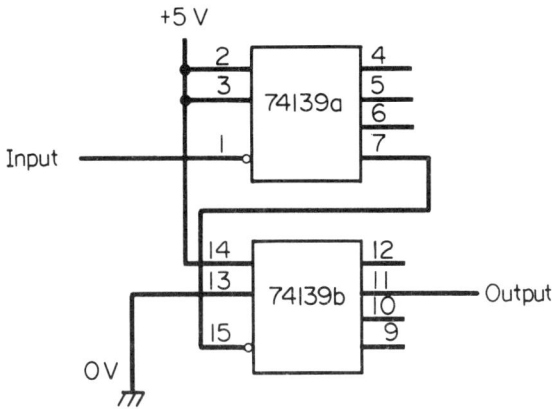

7. Design a logic system that outputs logic 1 to activate an audible alarm when an On/Off switch is set to On *and* an intruder detect switch is set *and* a timeswitch contact is closed, *or* a manual override test switch is closed.

8. State the Boolean expression for the output X of the following circuit in terms of the inputs A, B and C. Justify the following circuit description: "the output of the circuit is pulled low by either of inputs A or B, unless clamped when the input C is high".

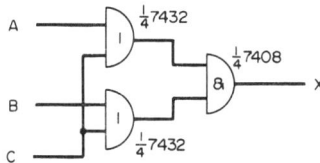

CHAPTER 6

Binary Numbers and Arithmetic

6.1 BINARY REPRESENTATION

The number system which man has long applied uses a base of ten, and is called the decimal or "denary" system. For example:

$$\text{denary } 728 = 728_{10} = 7 \times 10^2 + 2 \times 10^1 + 8 \times 10^0$$
$$= 700 \quad + \quad 20 \quad + \quad 8$$

Base of 10

Digital circuits (including computers) use the "binary" system to represent numbers. Binary representation uses the base of two, and only two symbols are used—0 and 1. For example:

$$\text{binary } 10101 = 10101_2 = 1 \times 2^4 + 0 \times 2^3 + 1 \times 2^2 + 0 \times 2^1$$
$$+ 1 \times 2^0$$

Base of 2

$$= 16 \quad + \quad 0 \quad + \quad 4 \quad + \quad 0$$
$$+ \quad 1$$
$$= 21_{10}$$

Thus denary 21 is represented by binary 10101. Each "bit" (binary digit) represents a power of 2, and so conversion from binary to denary is straightforward, as illustrated in the example above.

Conversion from denary to binary can also be achieved by synthesising a binary number using the powers of 2, as follows:

	2^5	2^4	2^3	2^2	2^1	2^0
	(32_{10})	(16_{10})	(8_{10})	(4_{10})	(2_{10})	(1_{10})
denary $39 = 39_{10} =$	1	0	0	1	1	1
$= 100111_2$						

An alternative method of denary to binary conversion involves a continual divide-by-two process, as follows for denary 19:

$$
\begin{array}{ll}
1 & 0 \leftarrow \text{fourth remainder} \\
2 \overline{\smash{)}2} & 0 \leftarrow \text{third remainder} \\
2 \overline{\smash{)}4} & 1 \leftarrow \text{second remainder} \\
2 \overline{\smash{)}9} & 1 \leftarrow \text{first remainder} \\
2 \overline{\smash{)}19} &
\end{array}
$$

Therefore $19_{10} = 10011_2$

The reader may like to confirm this result by reversing the process and performing binary to denary conversion (as shown above) on binary 10011. The answer should be denary 19.

6.2 HEXADECIMAL REPRESENTATION

Binary numbers tend to be long because many bits are required to represent even small denary numbers. Therefore binary numbers are frequently divided into groups of 4 bits, and each group is represented by a single character—these characters are 0, 1, 2, 3, 4, 5, 6, 7, 8, 9, A, B, C, D, E and F. These 16 symbols are called the "hexadecimal" code. The relationships between denary, binary and hexadecimal are shown in Table 6.1.

TABLE 6.1   NUMBER SYSTEMS

Denary	Binary	Hexadecimal
0	0000	0
1	0001	1
2	0010	2
3	0011	3
4	0100	4
5	0101	5
6	0110	6
7	0111	7
8	1000	8
9	1001	9
10	1010	A
11	1011	B
12	1100	C
13	1101	D
14	1110	E
15	1111	F

Consider the following 8-bit number as an example:

Binary: 0010   1101
Hexadecimal: 2   D

This can be expressed as $2D_{16}$ because hexadecimal numbers use a base of 16. An alternative way of expressing this number is therefore:

$$2 \times 16^1 + D(13_{10}) \times 16^0$$

Converting to denary: 32   +   13
$$= 45_{10}$$

Hexadecimal is a popular way of expressing binary numbers particularly with microcomputers, which commonly express numbers in 8-bit or 16-bit form. Notice that a group of 8 bits is frequently called a "byte".

6.3 BINARY ADDITION

The rules of binary addition are the same in principle as those for denary addition, as the following example illustrates:

$$14_{10} + 11_{10} = 1110_2 + 1011_2$$

Carry $\longrightarrow$ 1110
"Augend" $\longrightarrow$ 1110 $\longleftarrow$ 14_{10} +
"Addend" $\longrightarrow$ 1011 $\longleftarrow$ 11_{10}
$$\overline{11001} \longleftarrow 25_{10}$$

In this case of 4-bit addition a final fifth carry bit is required to represent the complete answer. A circuit arrangement to add two 4-bit numbers in this way is illustrated in Fig. 6.1. Provision is allowed for a Carry in bit, which may be required from a previous addition stage. For example, 8-bit addition can be achieved using two such devices connected in tandem, with one chip adding the least significant 4 bits and the other chip adding the most

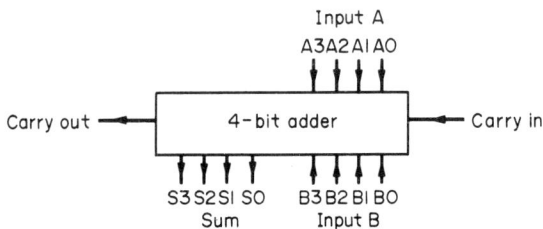

FIG. 6.1. 4-bit adder circuit.

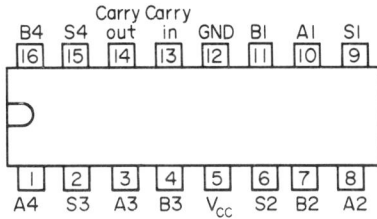

FIG. 6.2. 4-bit binary adder (7483).

significant 4 bits. The Carry in signal for the latter is fed from the Carry out signal from the former.

TTL chips are available to perform binary addition, e.g. the SN7483 4-bit binary full-adder shown in Fig. 6.2.

6.4 BINARY SUBTRACTION

Binary subtraction is performed as follows:

$$11_{10} - 6_{10} = 1011_2 - 0110_2$$

$$
\begin{array}{lll}
\text{Borrow} \longrightarrow & 0100 & \\
\text{"Minuend"} \longrightarrow & 1011 \longleftarrow 11_{10} \\
\text{"Subtrahend"} \longrightarrow & \underline{0110} \longleftarrow \underline{6_{10}} \\
& 0101 \longleftarrow 5_{10}
\end{array}
$$

Circuits to perform subtraction directly are not available. However, an adder circuit can be utilised if firstly the subtrahend is converted into its "two's complement" form, as shown in the following example for the numbers given above:

$$-6_{10} \text{ is created as follows:}$$
$$+6_{10} = 0110$$

Create "one's complement" (invert all bits): 1001
Create "two's complement" (add 1): 1010

therefore -6_{10} is represented by 1010

If this is added to the minuend:

$$
\begin{array}{ll}
11_{10} = & 1011 \\
-6_{10} = & \underline{1010} \\
& 10101 \longleftarrow \text{Difference } (11_{10} - 6_{10}) \\
& \uparrow
\end{array}
$$

Ignore final carry

The answer is seen to be 0101, as computed above.

The two's complement of any binary number is created by inverting all bits, and then adding 1, and is applied:

(a) to represent negative binary numbers, e.g. in computers (notice that the most significant bit is always set to 1),
(b) within the process of subtracting two positive binary numbers—the two's complement form of the subtrahend is created firstly before the numbers are added.

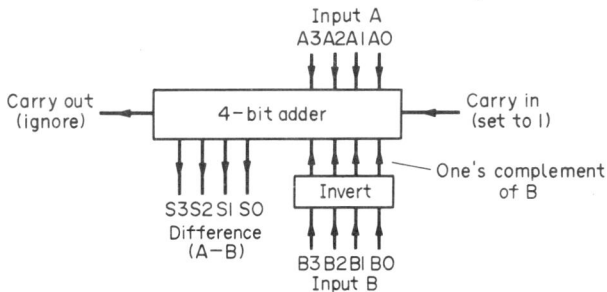

FIG. 6.3. 4-bit subtractor circuit.

Binary subtraction can be performed using the same 4-bit binary adder IC described in section 6.3. Figure 6.3 shows how the 4-bit subtrahend B is inverted to form the one's complement of B before entering the adder. The Carry in is set to 1 effectively to add 1 to B to create the two's complement form. This is added to A within the adder to create an output of A minus B.

6.5 BINARY MULTIPLICATION

Binary multiplication can be performed as follows:

$$13_{10} \times 9_{10} = 1101_2 \times 1001_2$$

$$
\begin{array}{r}
\text{"Multiplicand"} \longrightarrow 1101 \leftarrow 13_{10} \\
\text{"Multiplier"} \longrightarrow \underline{1001} \leftarrow 9_{10}
\end{array} \times
$$

$$
\text{Add four partial products}
\left\{
\begin{array}{r}
1101 \\
0000 \\
0000 \\
\underline{1101} \\
\hline
1110101 \leftarrow 117_{10}
\end{array}
\right.
$$

It can be seen that the binary multiplication process is effectively a series of add operations, with a left-shifted version of the multiplicand being added, or not added, depending on the 1 or 0 state of each bit in the multiplier. A 4-bit multiplier circuit can be constructed using three 4-bit adders. However whenever mathematical calculations of this type are required to be performed by an electronic system, it is customary to apply a computer solution (particularly with the advent of cheap microprocessor systems). The calculations are performed within the computer program, which can also perform an extensive range of other functions of course. Microcomputer systems are described in later chapters in this book.

6.6 BINARY DIVISION

Binary division is performed as follows:

$$15_{10} \div 4_{10} = 1111_2 \div 0100_2$$

$$\begin{array}{r} 11 \quad \longleftarrow \text{ Answer ("Quotient")} \\ \text{"Divisor"} \longrightarrow 0100 \overline{)1111} \quad \longleftarrow \text{ "Dividend"} \\ \underline{100} \\ 111 \\ \underline{100} \\ 11 \quad \longleftarrow \text{ Remainder} \end{array}$$

The answer is 11_2, remainder 11_2

i.e. 3_{10}, remainder 3_{10}

Thus binary division is effectively a series of subtraction processes.

Once again, although a binary divider circuit can be constructed using adders (connected as subtractors), it is customary to apply computers to such mathematical tasks.

BIBLIOGRAPHY

1. *Digital Techniques*, Barry G. Woollard, McGraw-Hill, 1983.

EXERCISES

1. Convert denary 41 to binary.
2. Convert binary 0101 1100 to denary.

3. Express binary 1000 1101 in hexadecimal.
4. Express hexadecimal A6 in binary.
5. Convert hexadecimal 2E to denary.
6. Convert denary 69 to hexadecimal.
7. Add binary 0011 0010 and 0001 1011.
8. Subtract binary 0010 0111 from 0111 1100.
9. Multiply binary 1101 by 0111. Express the answer as two hexadecimal characters.
10. Divide binary 0100 0001 by 0000 0101. Express the answer as two hexadecimal characters.
11. Draw the circuit diagram of an 8-bit adder constructed using two 7483 ICs.

CHAPTER 7

Analogue Circuits—the Op-amp Amplifier

7.1 THE BASIC OPERATIONAL AMPLIFIER

The operational amplifier, which is normally called the "op-amp", is a high gain dc amplifier (ac signals up to a limiting frequency are amplified as well). Unlike the previous digital circuits which have been examined in this book, the op-amp processes analogue (or continuous) signals. The extensive development of op-amps in integrated circuit form has produced cheap devices with greatly improved characteristics over their discrete component forerunners together with a wide range of applications.

The circuit symbol for an op-amp is shown in Fig. 7.1, together with the circuit of a simple transistor amplifier stage. The advantages which the op-amp offers over the latter are:

(a) far higher voltage gain—100,000 c.f. 80,
(b) rejects common mode noise (transistor amplifier provides little rejection),
(c) rejects supply voltage variations and effects of temperature variations,

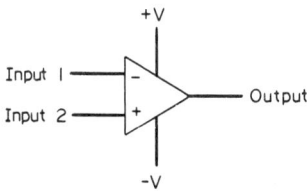

(a) Op-amp

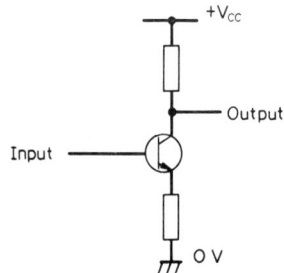

(b) Single transistor amplifier

FIG. 7.1. The op-amp and single transistor amplifiers.

65

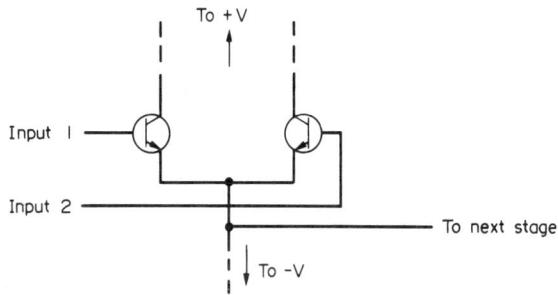

FIG. 7.2. Input stage to op-amp.

(d) higher input impedance—1 MΩ c.f. 20 KΩ,
(e) lower output impedance—150 Ω c.f. 10 KΩ.

A common mode noise signal is one which is applied to both input signal connections, and the method by which common mode signals (and supply voltage variations) are rejected in an op-amp is illustrated in Fig. 7.2. The two input signal connections feed two parallel transistors, such that this input stage amplifies only the *difference* between the input signals. Thus a common unwanted noise signal on both connections (this can occur when a transducer signal is amplified in a plant monitoring situation) is automatically cancelled and is not amplified. The input stage of the op-amp is often called a "differential amplifier" for this reason. Additionally if the supply voltage is poorly regulated and varies, this does not vary the gain of the amplifier, as occurs with a single transistor amplifier. Once again this effect is cancelled in the difference amplifier. A further benefit is that temperature variations, which can also vary the voltage gain of a single transistor amplifier, are similarly self-cancelling in the op-amp.

The extent of common mode noise rejection in an op-amp is normally expressed as the logarithm of the ratio of normal voltage gain to voltage gain of noise signal, and expressed in "decibels", as follows:

Common mode rejection ratio (CMRR) =

$$20 \log_{10} \frac{\text{Normal (differential) voltage gain (typically 100,000)}}{\text{Common mode voltage gain}}$$

where

$$\text{Common mode voltage gain} = \frac{\text{Differential output voltage (nearly 0)}}{\text{Common mode input voltage}}$$

Typically CMRR = 90 dB (decibels)

The op-amp is rarely applied in the open-loop configuration shown in Fig. 7.1. It is invariably applied with "feedback" components inserted between output and input, and these practical circuits are discussed in the subsequent sections.

7.2 THE INVERTING AMPLIFIER

Figure 7.3 shows an op-amp circuit connected as an inverting amplifier. Two resistors are applied to limit and regulate the voltage gain, and also to optimise the input impedance (as high as possible) and output impedance (as low as possible).

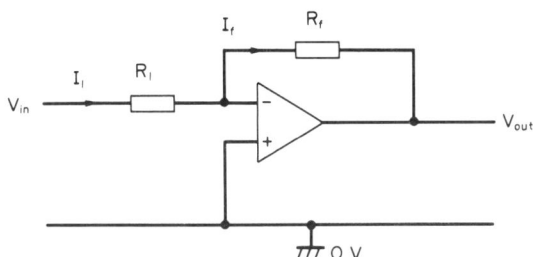

FIG. 7.3. Inverting amplifier.

The circuit analysis is straightforward because it can be assumed that the voltage at the inverting terminal $(-)$ is the same as that at the non-inverting terminal $(+)$. This is because the voltage gain of the device is so high and therefore the voltage difference between the two input terminals can be assumed to be negligible. The inverting terminal is said to be at "virtual earth". Additionally it can be assumed that no current flows into the device because the open loop input impedance is so high. Therefore:

$$V_{in} = I_1 R_1$$

and
$$V_{out} = -I_f R_f$$

Therefore the voltage gain is:

$$A_f = \frac{V_{out}}{V_{in}} = -\frac{I_f R_f}{I_1 R_1}$$

$$= -\frac{R_f}{R_1} \quad (\text{since } I_f = I_1)$$

This result is highly significant. It indicates that the voltage gain is independent of the open-loop gain of the device, which can be variable between devices, and is determined by the choice of two external resistors. The input impedance R_{in} to the overall circuit is:

$$R_{in} = \frac{V_{in}}{I_1} \quad \text{(due to the virtual earth)}$$

$$= R_1$$

Thus for typical resistor values of $R_f = 100 \text{ K}\Omega$ and $R_1 = 10 \text{ K}\Omega$:

$$\text{Voltage gain } A_f = 10$$
$$\text{Input impedance } R_{in} = 10 \text{ K}\Omega$$

The output impedance is virtually zero.

7.3 THE NON-INVERTING AMPLIFIER

Often it is required to amplify a signal without inversion and to produce a higher input impedance than the inverting amplifier offers in order to prevent electrical loading on the previous circuit. In this case the arrangement in Fig. 7.4 is applied.

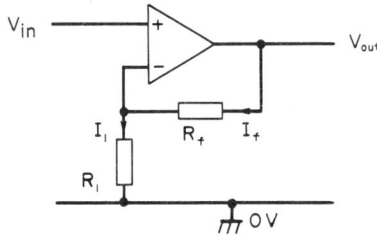

FIG. 7.4. Non-inverting amplifier.

The circuit analysis is once more based on the "virtual earth" principle, although in this case neither input is at earth but they can be considered to be at the same potential due to the high gain of the op-amp. The current through the two resistors is the same:

$$I_f = I_1$$

and therefore because voltage across R_1 is V_{in}:

$$\frac{V_{out} - V_{in}}{R_f} = \frac{V_{in}}{R_1}$$

SO
$$V_{out} = V_{in} \frac{R_f}{R_1} + V_{in}$$

Therefore voltage gain A_f is:

$$A_f = \frac{V_{out}}{V_{in}} = \frac{R_f}{R_1} + 1$$

This is almost identical to the voltage gain for an inverting amplifier, except for the $+1$, i.e. it is slightly higher.

It can further be proved that the overall input impedance

$$R_f = A . R_{in} \qquad \text{where } R_{in} \text{ is the input impedance of open-loop op-amp}$$

Thus the input impedance is virtually infinite (typically $10^{10}\ \Omega$).

7.4 SMALL DC VOLTAGE SIGNAL AMPLIFIER

In this and subsequent sections in this chapter we will examine amplifier circuits for different applications.

Figure 7.5 shows the simple arrangement which can be applied to amplify a small voltage signal from a transducer. The voltage transducer could be:

(a) a thermocouple, which generates a variable dc voltage (e.g. 0 to 50 millivolts) when ambient temperature varies,

(b) a solar cell, which generates a variable voltage in response to absorbed light,

(c) shaft speed indicator, e.g. a tachogenerator or an anemometer (rotating cup assembly which measures flow),

(d) a signal in an instrumentation system that has already been amplified but requires a further stage of amplification before application to a recording instrument, e.g. a servo in a pen recorder,

plus many others.

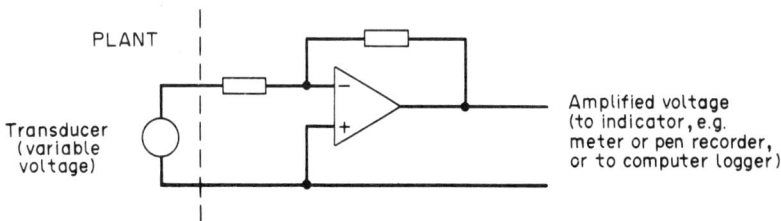

FIG. 7.5. Voltage transducer amplifier.

The dc amplification characteristic of the op-amp is clearly required in these instrumentation and plant measurement applications due to the slow-drifting (or even fixed level for long periods) nature of the measured signals.

7.5 BRIDGE SIGNAL AMPLIFIER

Most plant transducers are of the variable resistance type and not variable voltage as just described. Examples are shown in Fig. 7.6. In each case the transducer is simply a variable resistor.

The method by which the variable resistance value is converted into a voltage signal and subsequently amplified for connection to some form of indicator is shown in Fig. 7.7. Three resistors in the bridge are precision fixed-value components. As the resistance of the transducer, which forms the fourth arm of the bridge, varies a voltage unbalance signal is generated. This signal is amplified by the op-amp.

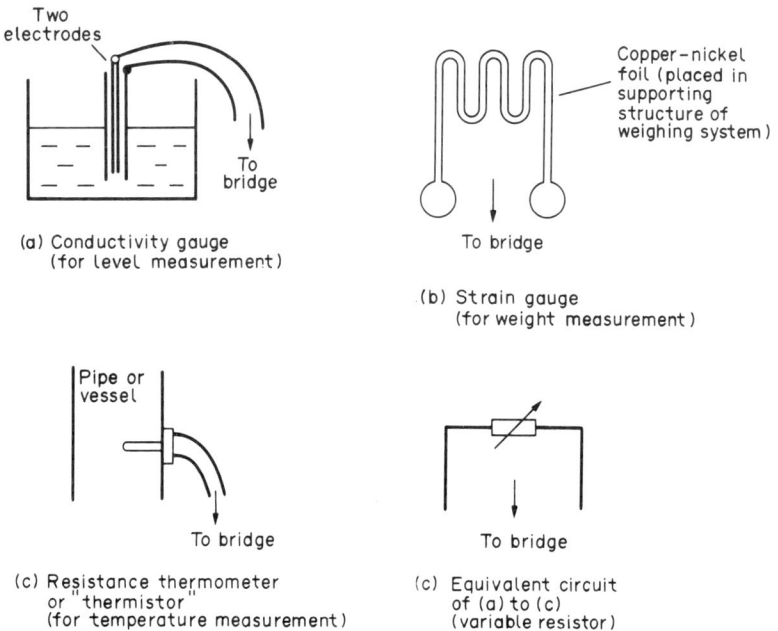

(a) Conductivity gauge
(for level measurement)

(b) Strain gauge
(for weight measurement)

(c) Resistance thermometer or "thermistor" (for temperature measurement)

(c) Equivalent circuit of (a) to (c) (variable resistor)

FIG. 7.6. Variable resistance transducers.

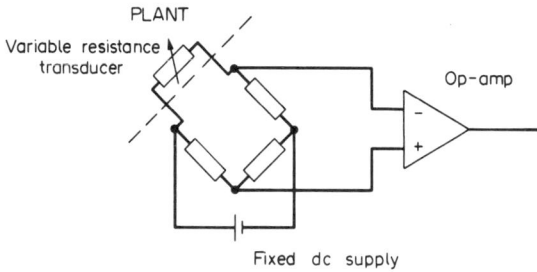

FIG. 7.7. Resistance transducer connection to Wheatstone bridge.

7.6 INSTRUMENT AMPLIFIER (MORE COMPLEX)

A more complex practical circuit that is commonly applied in instrumentation applications is shown in Fig. 7.8. This three op-amp configuration possesses the following features:

(a) extremely high input impedance to produce minimal loading on the feed circuit,
(b) excellent common mode rejection,
(c) low drift due to temperature variations,
(d) stable gain.

The overall specification exceeds that of a single op-amp circuit. The widespread application of this circuit has prompted chip manufacturers to offer it within a single IC package, e.g. the AD522 and the 725CN.

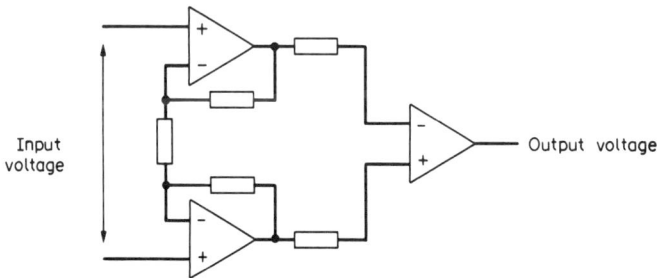

FIG. 7.8. Instrument amplifier using three op-amps.

7.7 AUDIO AMPLIFIER

All of the amplifier circuits described so far are basically dc amplifiers, but they also amplify ac signals up to a limiting frequency. However specific

ICM-F

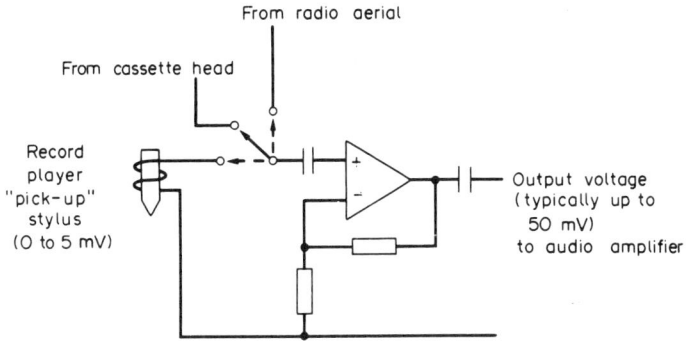

FIG. 7.9. Audio pre-amplifier.

op-amp circuits are used for audio amplification purposes, e.g. stereo amplifier and power amplifier to loudspeaker, and only ac frequencies over the audio bandwidth are amplified. A typical example is shown in Fig. 7.9. This pre-amplifier stage requires a high input impedance and the other characteristics that an op-amp offers. The capacitors block dc but pass ac signals. In practice other components (resistors and capacitors) are included in the feedback path to provide "equalisation', i.e. different gain/frequency characteristics for the type of signal which is amplified (radio, record pick-up or magnetic cassette head).

This input stage typically passes to a second op-amp stage, which provides further voltage gain as well as tone (treble, bass, "scratch" rejection) control, as shown in Fig. 7.10. The treble and bass controls accentuate high and low frequency signals respectively. Although not shown in this diagram an RC filter can often be switched in to the circuit to eliminate "scratch"

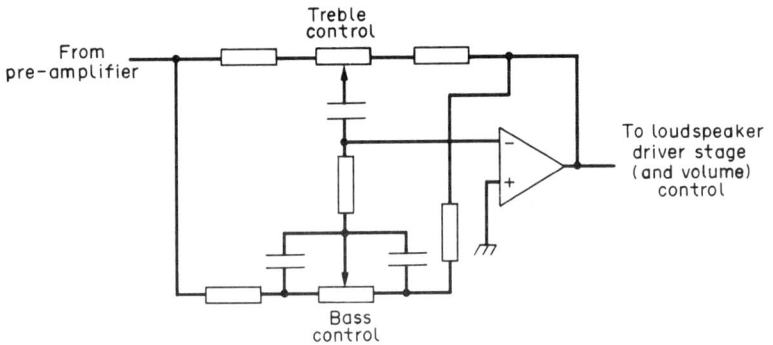

FIG. 7.10. Audio tone control amplifier stage.

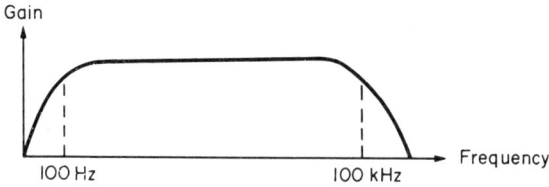

FIG. 7.11. Frequency response of audio amplifier.

noise. This simply serves to reduce the high frequency gain of the overall response of the circuit, as shown in Fig. 7.11.

Op-amp amplifiers cannot amplify radio frequency signals and television signals, but they are applied widely for audio systems (up to 100 kHz).

7.8 REAL OP-AMPS

A large variety of op-amp ICs is available. The first widely applied device was the 709, but this has been largely superseded by the 741. The 741 is virtually an industry standard because it offers an excellent all-round performance at a low price; its cheapness is a measure of its popularity.

Figure 7.12 shows the pin layout and circuit connection of a 741 (full description is SN72741). The purpose of the offset null potentiometer is to

(a) Package

Offset null	1		8	NC (Not connected)
Input (-)	2		7	V+
Input (+)	3	741	6	Output
-V	4		5	Offset Null

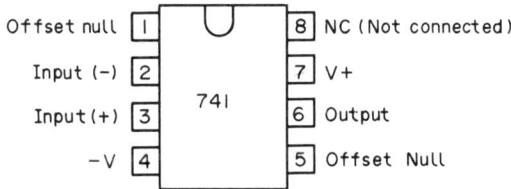

(b) Circuit connection (with offset null)

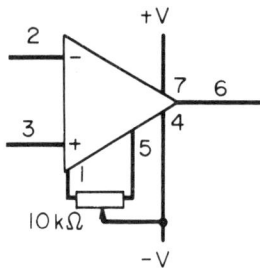

Normally:
+V = +15 V
-V = -15 V
from dual power supply

FIG. 7.12. 741 op-amp.

eliminate output offset voltage, which can be explained as follows. Even when there is no input voltage, an op-amp exhibits a small voltage between the inverting and non-inverting terminals. This is amplified (by a factor determined by the external resistors—see Fig. 7.3 and Fig. 7.4) and appears at the output. This output error signal could be 0.1 V or more. It is isolated from subsequent stages in an ac amplifier circuit by a blocking capacitor, but the 10 KΩ potentiometer shown in Fig. 7.12(b) is required for dc connection. Both inputs should be connected to ground (0 V) whilst this potentiometer is adjusted to cancel the output offset voltage before the circuit is used.

The 741 offers additional features of short-circuit protection on the output circuit (this is particularly important because the output impedance is so low) and internal frequency compensation (a 30pF capacitor is fabricated within the chip). This latter feature is required to prevent unwanted oscillations occurring when some methods of feedback are applied. Other op-amps frequently do not provide internal frequency compensation. This may be an advantage if it is required to amplify high frequencies because a selected external capacitor value can enhance the frequency response.

It is not sensible to quote the frequency response of a particular op-amp because this parameter is governed by the complete circuit arrangement and the actual values of the input and feedback resistors.

In any feedback amplifier:

$$\text{Gain} \times \text{Bandwidth} = \text{Constant}$$

and so an increased frequency response (bandwidth) can be obtained by reducing the amplifier gain. The parameter that is quoted for an open-loop op-amp to indicate frequency response is the "slew rate", which is defined as the maximum rate at which the output voltage can change. For example, if

$$\text{slew rate} = 1 \text{ V}/\mu s$$

and the output changes by 5 V, time taken is 5 μs.

A range of MOS FET-input devices has been added to the op-amp range. They feature higher input impedances than the traditional bipolar op-amps, and are applied with high-impedance transducers such as radiation detectors, photomultipliers and pH cells. Typical examples are the 3140E, which is pin-compatible with the 741, and the AD540J. A further development to the FET-input devices is the CMOS op-amp range, which offers reduced power consumption, e.g. 250 mW c.f. 500 mW for bipolar and FET devices. CMOS op-amps are particularly suited to battery-driven systems, and will operate off a single power supply, e.g. a +8 V and 0 V nickel-cadmium cell. A typical device is the 7611, which again is pin-compatible with the 741.

BIBLIOGRAPHY

1. *Using Digital and Analogue Integrated Circuits*, L. W. Shaklette and H. A. Ashworth, Prentice/Hall, 1978.
2. *Analog Data Manual 1982*, Mullard, Signetics, 1982.

EXERCISES

1. Suggest two advantages which an op-amp in IC form posesses over its discrete component forerunner, which consisted of typically 20 transistor circuits mounted on a board.
2. Why is "common mode rejection" important when amplifying plant instrumentation signals, and how is it achieved in an op-amp?
3. What is the "virtual earth" principle that is applied in op-amp circuit analysis?
4. Design an inverting op-amp circuit that *attenuates* the input signal by a factor of 10 and has an input impedance of 100 kΩ.
5. Calculate the voltage gain of a non-inverting amplifier circuit in which the feedback resistor is 47 kΩ and the input resistor is 10 kΩ.
6. Sketch an amplifier circuit, based on an op-amp, to process a signal from a transducer which produces a variable electrical resistance when the pressure in a gas pipe varies.
7. What is "equalisation" in an audio amplifier?
8. What is a typical upper limit to the frequency response of an op-amp amplifier, and how can it be varied?
9. What is output offset voltage in an op-amp, and how is it eliminated in the 741?
10. What particular advantage is offered over the 741 op-amp by each of the following?
    (a) MOS FET-input op-amp, e.g. 3140E.
    (b) CMOS op-amp, e.g. 7611.

CHAPTER 8

Analogue Circuits—Other Applications of the Op-amp

8.1 VOLTAGE FOLLOWER

The simplest op-amp closed-loop circuit is shown in Fig. 8.1. This circuit acts as a buffer, or a "voltage follower". Assuming that the input signal connections ($+$) and ($-$) are at the same potential due to the high open-loop gain of the device, then clearly

$$V_{out} = V_{in}$$

Thus the closed-loop gain is 1. The value of the circuit is that the input impedance is very large and the output impedance is very low, and the circuit is used for impedance matching.

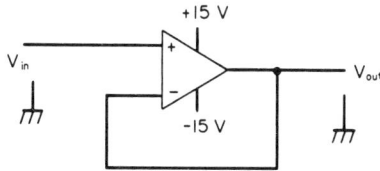

FIG. 8.1. Voltage follower.

8.2 COMPARATOR

An op-amp connected in the open-loop arrangement shown in Fig. 8.2 acts as a comparator if one of the inputs is held at a reference voltage. When $V_{in} > V_{ref}$ the output is driven into negative saturation, and when $V_{in} < V_{ref}$ the output is at positive saturation. If these saturation levels are clamped to 0 V and $+5$ V they are suitable for driving digital ICs.

76

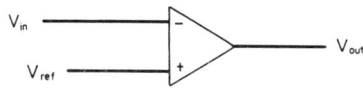

FIG. 8.2. Simple comparator.

A common practical comparator circuit is the zero-crossing detector shown in Fig. 8.3. At any time one of the two zener diodes is at its reverse breakdown voltage (V_z) depending on the polarity of V_{in}. The output voltage is therefore at

$$+V_Z \quad \text{or} \quad -V_Z$$

and is said to be "clamped". This clamping effect improves the high frequency response of the comparator.

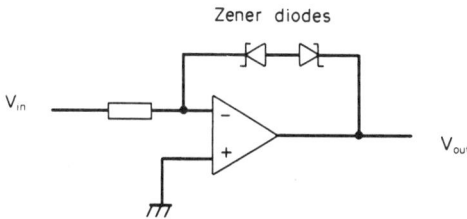

FIG. 8.3. Zero-crossing detector.

Another popular comparator circuit that indicates when an input voltage has reached a reference value is shown in Fig. 8.4. In this case V_{out} switches from $+V_Z$ to $-V_Z$ when V_{in} increases past

$$\frac{R_{in}}{R_{ref}} \times V_{ref}$$

When a comparator feeds into a digital circuit the generation of TTL logic levels can be accomplished as follows:

(a) use a single-supply op-amp, e.g. the 3140E which was mentioned in

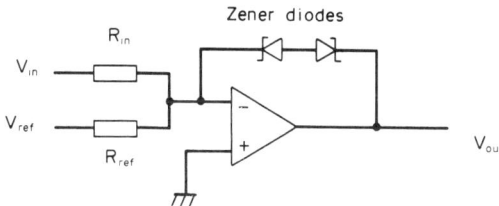

FIG. 8.4. Comparator with zener diode clamp.

section 7.7, rather than the conventional dual supply (+15 V and −15 V) device.
(b) use a special-function op-amp comparator IC, e.g. the 311, which generates TTL signals.

8.3 SUMMER (ADDER)

A circuit that sums, or adds, two or more voltage signals is shown in Fig. 8.5. The circuit analysis is based upon the assumptions that the voltage on the (−) input terminal is zero and that no current flows into the op-amp. Then

$$I_1 + I_2 + I_3 = I_f$$

Therefore

$$\frac{V_1}{R_1} + \frac{V_2}{R_2} + \frac{V_3}{R_3} = -\frac{V_{out}}{R_f}$$

and so

$$V_{out} = -R_f\left(\frac{V_1}{R_1} + \frac{V_2}{R_2} + \frac{V_3}{R_3}\right)$$

which is a summing function.

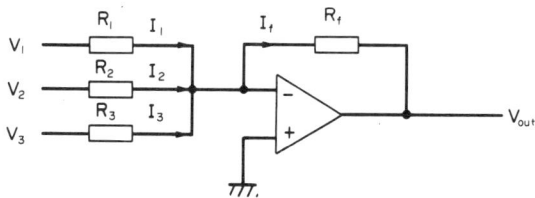

FIG. 8.5. Summer (adder) circuit.

The input voltages can be dc or ac. A simple averaging circuit, which generates the mean of two input voltages, could use:

$$R_f = 10\,K\Omega, \qquad R_1 = 20\,K\Omega, \qquad R_2 = 20\,K\Omega \quad \text{and no } R_3$$

Therefore $\qquad V_{out} = -\tfrac{1}{2}(V_1 + V_2)$

8.4 DIFFERENCE AMPLIFIER (SUBTRACTOR)

The basic adder circuit can be converted into a subtractor if one of the input voltages is fed into the non-inverting input (+), as shown in Fig. 8.6.

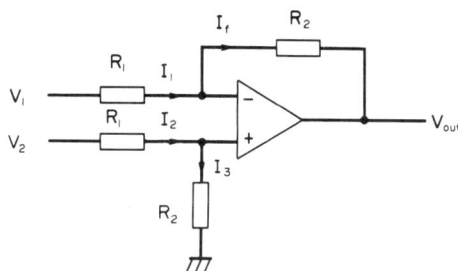

FIG. 8.6. Difference (subtractor) circuit.

Using the assumptions that:

$$I_1 = I_f \quad \text{and} \quad I_2 = I_3$$

a straightforward but lengthy analysis leads to:

$$V_{out} = \frac{R_2}{R_1}(V_2 - V_1)$$

which is a subtraction, or difference, function.

A variation of this basic subtractor circuit can be used for a different application—to convert a floating signal to a grounded signal. Consider Fig. 8.7, in which an ungrounded, or "floating", signal is connected to the two inputs of a difference or subtractor circuit. The overall voltage gain is given by:

$$V_{out} = \frac{2\ M\Omega}{2\ M\Omega}(V_{in})$$

so that voltage gain is 1.

The resistor values are chosen deliberately to be high.

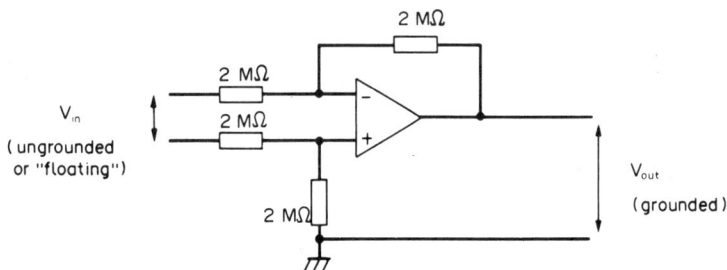

FIG. 8.7. Conversion of floating signal to grounded signal using difference amplifier.

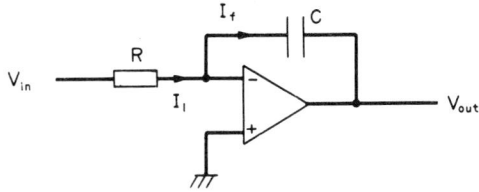

FIG. 8.8. Integrator.

8.5 INTEGRATOR

A common application of an op-amp is in an integrating circuit shown in Fig. 8.8. Making the usual assumptions about the inverting terminal (no voltage, no current into the op-amp) the voltage across the capacitor can be expressed as:

$$V_{out} = -\frac{1}{C} \int I_f \, . \, dt$$

Proof:

$$V = \frac{Q}{C} \quad \text{for a capacitor} \qquad Q = \text{charge across capacitor}$$

Differentiating: $\dfrac{dV}{dT} = \dfrac{1}{C} \dfrac{dQ}{dt} = \dfrac{I}{C}$ I = rate of change of charge

Integrating: $V = \dfrac{1}{C} \int I \, . \, dt$ □

Therefore $V_{out} = -\dfrac{1}{C} \int \dfrac{V_{in}}{R} \, . \, dt$ since $I_f = I_1 = \dfrac{V_{in}}{R}$

$$= -\frac{1}{CR} \int V_{in} \, . \, dt$$

Thus the output voltage is the integral (or time sum) of the input voltage, multiplied by a constant.

The effect of this circuit on a variety of input voltage waveforms is illustrated in Fig. 8.9. Notice that the output voltage is inverted; inversion is not shown in (a) for clarity.

The principal difficulty with this circuit is that a small input offset voltage (described in section 7.7) causes the output voltage to ramp to the saturation state in the manner shown in Fig. 8.9(c) The solution is to place a switch (typically a FET) across the capacitor, such that the switch is only opened during the actual integration period.

(a) Sinewave input (inversion on V_{out} not shown)

(b) Squarewave input

(c) Step input

(d) Squarewave input (positive-going only)

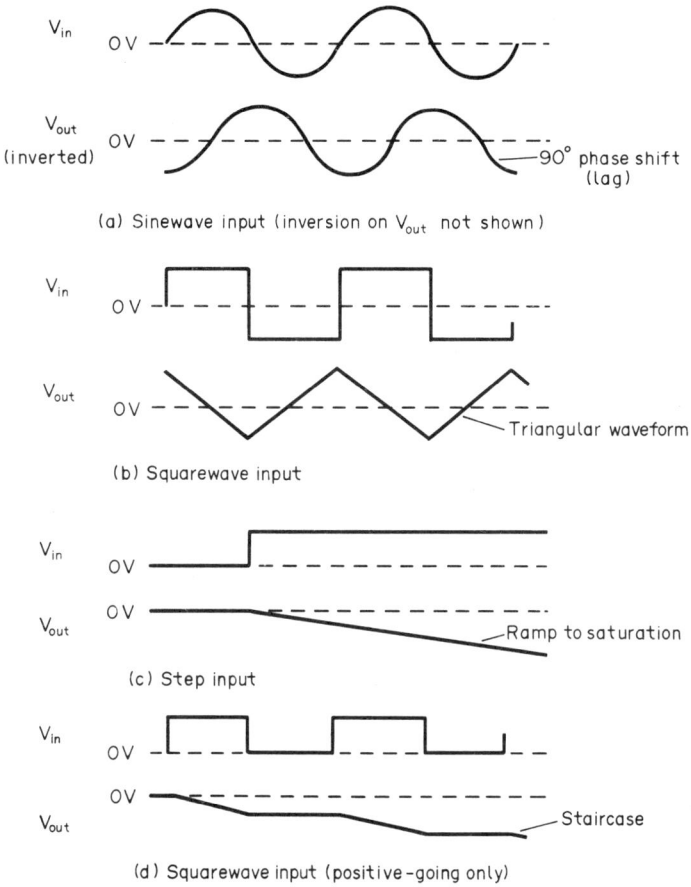

FIG. 8.9. Integrator voltage waveforms.

8.6 DIFFERENTIATOR

A differentiator circuit is shown in Fig. 8.10. For this configuration:

$$V_{out} = -I_f R = -I_1 R$$

$$= -\left(C \frac{dV_{in}}{dt}\right) R$$

Proof: As above, using Q = CV for a capacitor

$$I = \frac{dQ}{dt} = C \frac{dV}{dt} \qquad \square$$

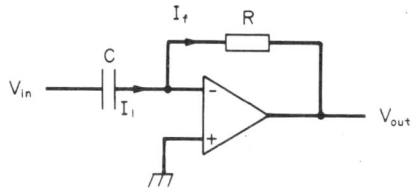

FIG. 8.10. Differentiator.

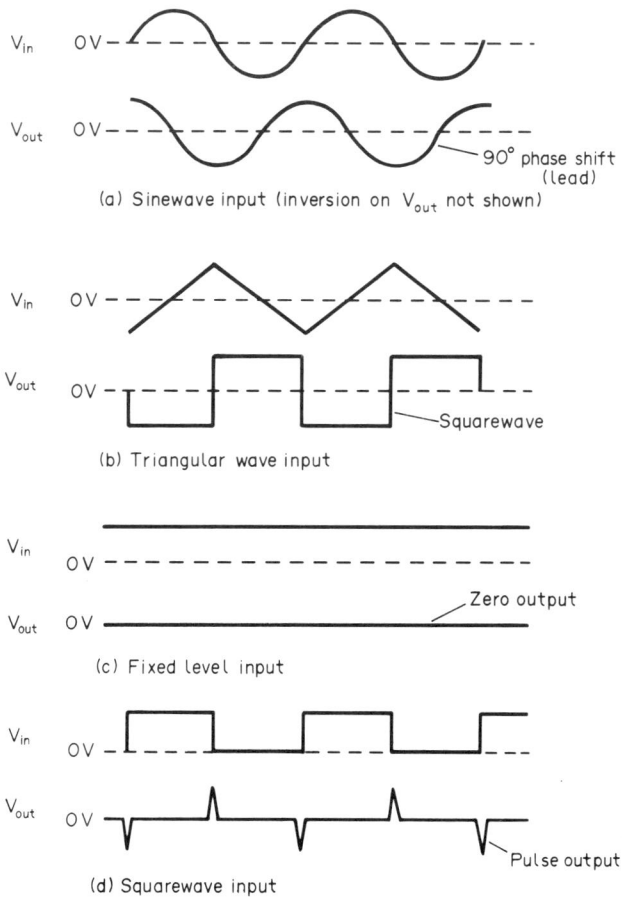

(a) Sinewave input (inversion on V_{out} not shown)

(b) Triangular wave input

(c) Fixed level input

(d) Squarewave input

FIG. 8.11. Differentiator voltage waveforms.

Therefore $$V_{out} = -CR \frac{dV_{in}}{dt}$$

Thus the output voltage is the differential (or rate of change) of the input voltage, multiplied by a constant.

The effect of this circuit on a variety of input voltage waveforms is illustrated in Fig. 8.11. Notice that the output voltage is inverted; inversion is not shown in (a) for clarity. The action of the differentiator therefore is that an output voltage is produced only if the input voltage changes, and the output voltage is proportional to the rate of change of the input voltage. In (b) the output voltage is at a fixed level when the input voltage ramps upwards, and at a level of the opposite polarity when the input ramps downwards. In (d) the rising and falling edges of the input waveform are of almost infinite slope, and so the output waveform consists of narrow pulses (perhaps of saturation magnitude).

8.7 SCHMITT TRIGGER

A Schmitt trigger circuit is employed when it is required to produce a voltage transition when an input voltage exceeds a threshold value, and reverses the transition only when the input voltage falls substantially below that threshold level. Thus it is a form of comparator circuit, but it possesses "hysteresis", as illustrated in Fig. 8.12.

$$\text{Hysteresis} = V_{on} - V_{off}$$

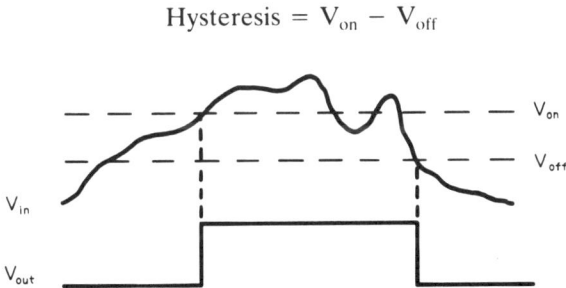

FIG. 8.12. Schmitt trigger response.

An op-amp Schmitt trigger circuit is shown in Fig. 8.13. Notice that feedback is applied to the non-inverting terminal. The output voltage switches to negative saturation when the input voltage exceeds V_{on}. The output voltage switches back to positive saturation when the input voltage subsequently falls below V_{off}. Circuit analysis can prove that:

$$V_{on} = \frac{V_{max}R_1 + V_{ref}R_2}{R_1 + R_2}$$

V_{max} and V_{min} are the saturation voltages

$$V_{off} = \frac{V_{min}R_1 + V_{ref}R_2}{R_1 + R_2}$$

(+ and − supply voltages)

Thus a hysteresis of 10 V can be produced if:

$$R_1 = 10\,K\Omega, \qquad R_2 = 20\,K\Omega,$$
$$V_{ref} = +5\,V, \qquad V_{max} = +15\,V, \qquad V_{min} = -15\,V$$

A Schmitt trigger circuit is offered in the TTL 7400 digital IC series, and this is described in section 5.3. The hysteresis is much less (typically 0.8 V) in the 7400 series however.

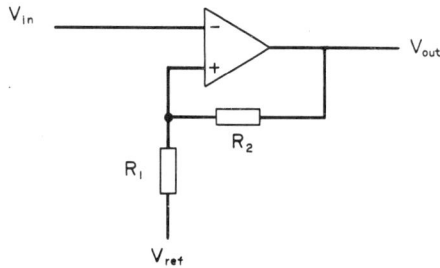

FIG. 8.13. Op-amp Schmitt trigger circuit.

8.8 ANALOGUE MULTIPLIERS AND DIVIDERS

The logarithmic amplifier in Fig. 8.14 makes use of the fact that the voltage/current relationship for the diode is highly non-linear. This leads to the following relationship:

$$V_{out} = K \log_{10} V_{in} \qquad \text{where } K = \text{constant}$$

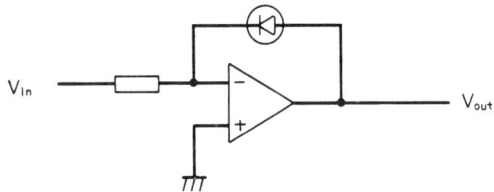

FIG. 8.14. Logarithmic amplifier.

This amplifier therefore can be applied when it is required to compress the lower end of the input voltage range; the output voltage can then feed a non-linear indicator which accentuates the upper part of the signal range. An alternative application is in a multiplier or divider circuit—these are described later in this section. Note that a transistor is used in place of the diode in some applications.

The antilogarithmic amplifier in Fig. 8.15 generates the following relationship:

$$V_{out} = K \text{ antilog}_{10} V_{in} \qquad \text{where } K = \text{constant}$$

Both of these amplifier circuits are offered by some manufacturers in a single package—there are no external components. The value of these circuits is that multiplication of analogue signals can be achieved. Consider the following standard mathematical relationship:

$$V_1 \times V_2 = \text{antilog}_{10} (\log_{10} V_1 + \log_{10} V_2)$$

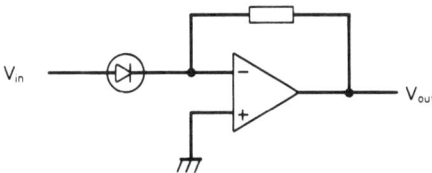

FIG. 8.15. Antilogarithmic amplifier.

This relationship is generated by the circuit shown in Fig. 8.16. The two logarithmic amplifiers are followed by a summer (see section 8.3) and then an antilogarithmic amplifier.

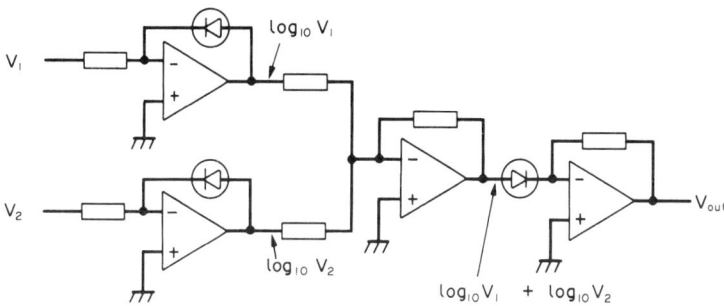

FIG. 8.16. Analogue multiplier.

Similarly division of analogue signals can be performed using the following relationship:

$$\frac{V_1}{V_2} = \text{antilog}_{10}\,(\log_{10} V_1 - \log_{10} V_2)$$

In this case the summer in Fig. 8.16 is replaced by a subtractor (see section 8.4).

The drawback with the multiplier and divider circuits just described is that they are only "1-quadrant" devices, i.e. they operate on signals which are both of fixed polarity, e.g. both are positive voltages. If it is required to multiply and divide two signals, both of which can be of positive or negative polarity, then a "4-quadrant" circuit is required. Such a device is available in IC form, as shown in Fig. 8.17. The output is a scaled (by a factor of 10) version of the product of the two input signals to prevent saturation. 4-quadrant multipliers tend to be expensive; a typical device is the AD534JH. They are not as accurate as log/antilog multipliers at low signal levels, but they possess better speed and bandwidth.

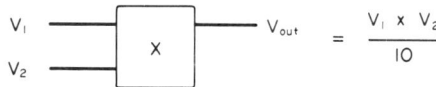

FIG. 8.17. Multipler IC.

Figure 8.18 shows how a 4-quadrant multiplier can be used in the feedback path of an op-amp to produce a divide function. Making the normal assumptions about an op-amp:

$$I = I_f$$

Therefore

$$V_1 = -V_M \qquad \text{(resistors are of same value, and so voltage across them must be the same)}$$

$$= -\frac{V_2 \times V_{out}}{10}$$

So
$$V_{out} = -10 \times \frac{V_1}{V_2}$$

Therefore one input voltage is divided by the other (with a scaling factor of 10).

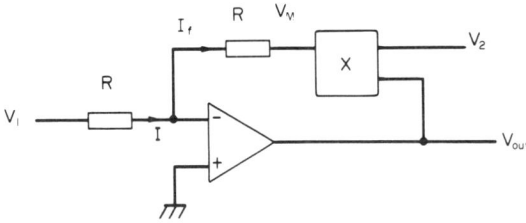

FIG. 8.18. Divide circuit.

8.9 SQUARE AND SQUARE ROOT GENERATION

A square function is generated by simply multiplying a signal by itself—see Fig. 8.19.

$$V_{out} = \frac{V_{in} \times V_{in}}{10} = \frac{V_{in}^2}{10}$$

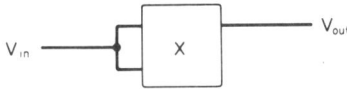

FIG. 8.19. Square (using multiplier).

A square root operation can be performed by inserting a square function circuit in the feedback path of an op-amp, as shown in Fig. 8.20. As stated in section 8.8 (divide function):

$$V_{in} = -V_M$$

Therfore

$$V_{in} = -\frac{V_{out}^2}{10}$$

So

$$V_{out} = \sqrt{10\, V_{in}}$$

Thus the output signal is proportional to the square root of the input.

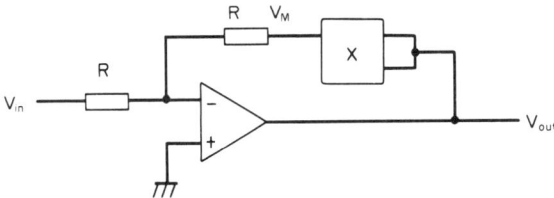

FIG. 8.20. Square root (using multiplier).

ICM-G

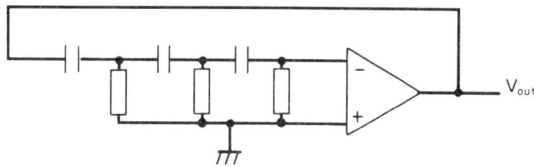

FIG. 8.21. Phase-shift oscillator.

8.10 OSCILLATORS

A feedback amplifier breaks into oscillation if the output signal adds to the input signal at any frequency when the total loop gain is greater than 1. The op-amp amplifier in Fig. 8.21 applies feedback to the inverting terminal. The RC network introduces at total of 180° phase shift, the op-amp itself adds 180° phase shift, and so the total phase shift is 360°. This positive feedback induces oscillations in the overall amplifier circuit, and a sinewave voltage is produced at the output.

A second type of oscillator, which produces a squarewave, is shown in Fig. 8.22. This circuit is similar to the Schmitt trigger (Fig. 8.13), except that an RC circuit is added. The output voltage continually switches from positive saturation to negative saturation. Assume that the output is initially at positive saturation. Current flows through R_f to charge the capacitor so that the voltage on the inverting terminal increases towards positive saturation. When it exceeds the voltage on the non-inverting terminal, as follows:

$$\frac{R_1}{(R_1 + R_2)} \times V_{out}$$

the device operates as a comparator, and the output switches to negative saturation. The process then repeats in the reverse direction to produce a squarewave output. The frequency of the generated pulse stream is governed by the choice of R and C values.

An alternative name for this circuit is an "astable multivibrator".

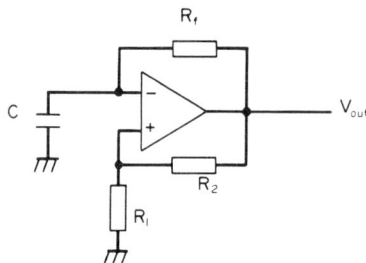

FIG. 8.22. Free running oscillator (astable multivibrator).

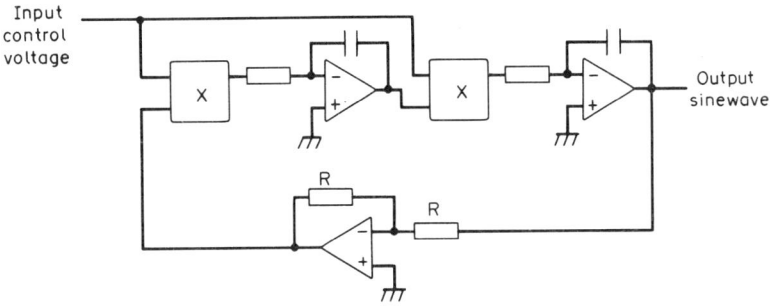

FIG. 8.23. Voltage controlled oscillator.

An oscillator circuit, in which the output frequency is varied by an input control voltage, is illustrated in Fig. 8.23. Each integrator gives 90° phase shift, whilst the unity gain inverter in the overall feedback path gives 180° phase shift in order to sustain oscillations. The control voltage enters the two multipliers and serves to vary the time constant of the integrator circuits. Typically a control voltage range of 1 V to 10 V gives a frequency range of 100 Hz to 1000 Hz.

8.11 POWER SUPPLY REGULATION

The standard dc power supply circuit as shown in Fig. 8.24, followed by an op-amp voltage follower, produces a regulated dc supply. Although the voltage across the zener diode is fairly well stabilised, it can vary slightly when the current through the zener diode alters due to variations in the load. The voltage follower presents such a high impedance to the zener diode that the zener current is held constant, and so the output voltage is fixed.

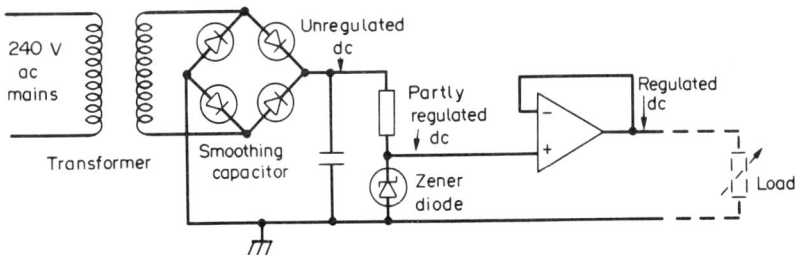

FIG. 8.24. Op-amp regulated power supply.

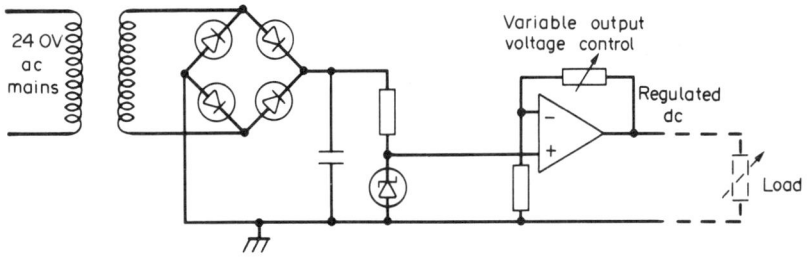

FIG. 8.25. Voltage-controlled regulated power supply.

A variable power supply can be created by replacing the voltage follower with a non-inverting amplifier, as shown in Fig. 8.25.

8.12 FILTERS

A filter is a circuit that passes ac signals of selected frequencies. Different resistor and capacitor circuits can be employed with an op-amp amplifier to produce:

(a) a low pass filter, i.e. passes only low frequency signals,
(b) a high pass filter, i.e. passes only high frequency signals,
(c) a bandpass filter, i.e. passes a restricted range of frequencies as shown in Fig. 8.26.

The cut-off frequencies for (a) and (b) are given by:

$$\text{Low pass cut-off frequency} = \frac{1}{R\sqrt{C_1 C_2}}$$

$$\text{High pass cut-off frequency} = \frac{1}{C\sqrt{R_1 R_2}}$$

The mid-band frequency of (c) is given by:

$$\text{Mid-band frequency} = \frac{\sqrt{2}}{RC}$$

and the "Q factor" is:

$$Q = \frac{\text{Mid-band frequency}}{\text{Bandwidth}} = \frac{\sqrt{2} R_1}{4R_1 - R_2}$$

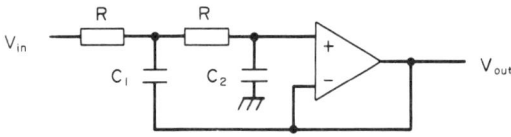

(a) Low pass filter

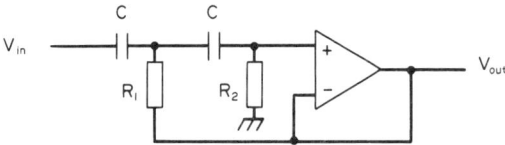

(b) High pass filter

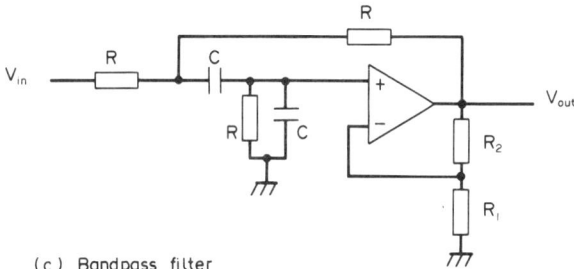

(c) Bandpass filter

FIG. 8.26. Filters.

BIBLIOGRAPHY

1. *Digital-Integrated-Circuit, Operational-Amplifier and Optoelectronic Circuit Design*, Bryan Morris, McGraw-Hill, 1976.
2. *Designing with Operational Amplifiers*, Jerald G. Graeme, Burr-Brown, 1977.
3. *An Introduction to Operational Amplifiers with Linear IC Applications*, Luces M. Faulkenberry, Wiley, 1982.

EXERCISES

1. Design an op-amp circuit that produces an output of +15 V when an input voltage is below a reference voltage level of +3 V, and switches to −15 V when the input voltage exceeds this level.

2. Design an op-amp circuit that sums three input voltages in increasing binary weighted proportions.

3. The following voltage waveform is connected to an op-amp circuit:

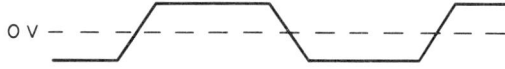

Sketch the output waveforms if the op-amp is connected as:

(a) an integrator,
(b) a differentiator.

4. Sketch an op-amp circuit to generate a signal which feeds to an alarm indicator circuit when the rate of change of a plant transducer voltage signal exceeds a present value.

5. Calculate the output voltage of the following circuit:

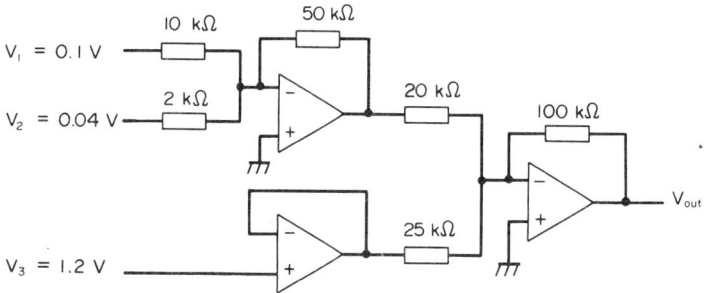

6. Sketch a circuit using logarithmic/antilogarithmic amplifiers to multiply four positive voltage signals.

7. An instrumentation application is required to produce a signal which is derived from two flow measurements P and Q using the following formula:

$$\text{Output} = P + \frac{dP}{dt} + Q^2$$

Draw a circuit using op-amps to perform this function.

8. Repeat 7 for the following expression:

$$\text{Output} = 2\sqrt{\frac{P \times D}{T}}$$

for the three inputs P (pressure), D (differential pressure) and T (temperature)

9. We have encountered in this book three different ICs which can be used to produce astable multivibrator circuits by the addition of external resistors and capacitors. Name them.

10. Why is an op-amp sometimes used in a dc power supply?

11. Design an op-amp circuit that amplifies ac signals up to a frequency of 5 MHz.

CHAPTER 9

Signal Mode Conversion

9.1 ANALOGUE TO DIGITAL (A/D) CONVERSION

Analogue to digital conversion is required when it is necessary to process an analogue voltage signal, e.g. an instrumentation signal for process flow, temperature, pressure, etc., within a digital electronic system. The latter is normally a computer.

The normal pin functions which are applied with an A/D converter IC are summarised in Fig. 9.1. When Start Conversion signal is set the device samples the input analogue voltage, and after a fixed number of clock pulses the digital representation of the input signal is set on the ten output pins and the Conversion Complete signal is set. This example gives a 10-bit output digital representation of the analogue input signal, i.e.

$$\text{Resolution} = 1 \text{ in } 2^{10} = 1 \text{ in } 1024$$
$$= 0.1\%$$

Alternative ICs give 8-bit and 12-bit outputs, and the accuracy of the

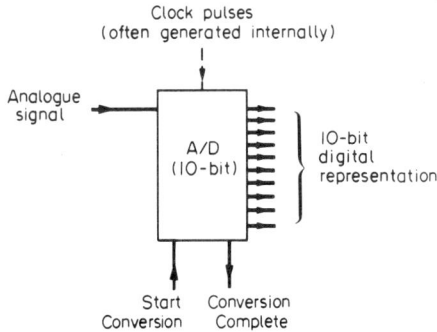

Fig. 9.1. A/D converter pin functions.

analogue signal and the required resolution of the converter chip for a particular application determines whether an 8, 10 or 12-bit device is chosen.

There are two different conversion techniques applied with A/D converter ICs:

(a) *Successive Approximation Technique*

In this arrangement successively reducing proportions of a reference voltage are added together and compared with the analogue input voltage, as shown in Fig. 9.2. In this example only 4 bits are generated for simplicity. Each resistor switches in a binary weighted fraction of a reference voltage.

The conversion sequence is as follows. When Start Conversion is set, the first Clock pulse causes resistor R to be switched into circuit. This introduces one-half of the reference voltage, and this voltage is effectively compared with the analogue input voltage. If the latter is the greater, then the resistor is left switched in; otherwise it is switched out. On the next Clock pulse resistor 2R is switched into circuit, and this introduces one-quarter of the reference voltage. The sum of the generated voltage components is then again compared with the analogue input voltage to determine whether 2R is to be switched out or left in circuit. The process repeats until all four voltage comparisons are made and a combination of resistors is left switched into circuit. The register contains a bit pattern reflecting the state of these resistor switches, and this bit pattern is presented on the digital output pins together with Conversion Complete after the last comparison process.

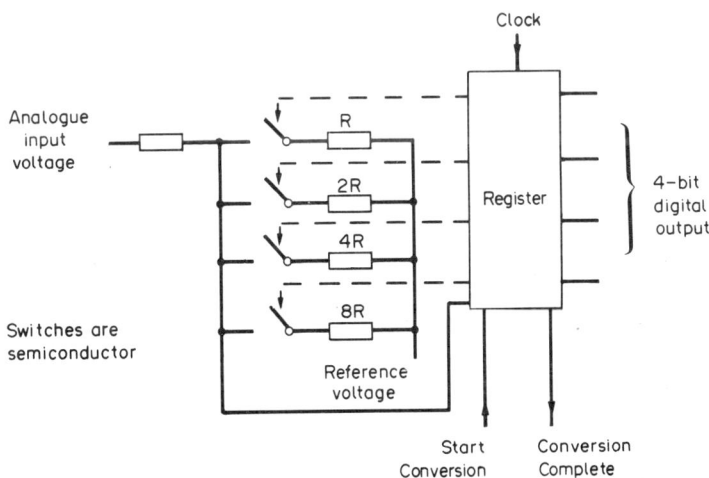

FIG. 9.2. Successive approximation A/D conversion process.

Conversion time is determined by the Clock rate, but is typically 15 μs (microseconds) for a 10-bit converter. This high speed makes it the most popular A/D converter.

(b) *Dual Ramp Technique*

In this technique illustrated in Fig. 9.3, a voltage integrator is firstly ramped up for a fixed time with the analogue input voltage, and secondly ramped down with a reference voltage of opposite polarity. Notice that the positive slope is steeper for the higher analogue input voltage. The negative slope is fixed, so that the integrator output takes a longer time to reach zero for the higher analogue input voltage. Pulses are gated into a counter during the negative slope period until terminated by the zero crossing detector. Therefore the higher the analogue input voltage, the greater is the count achieved.

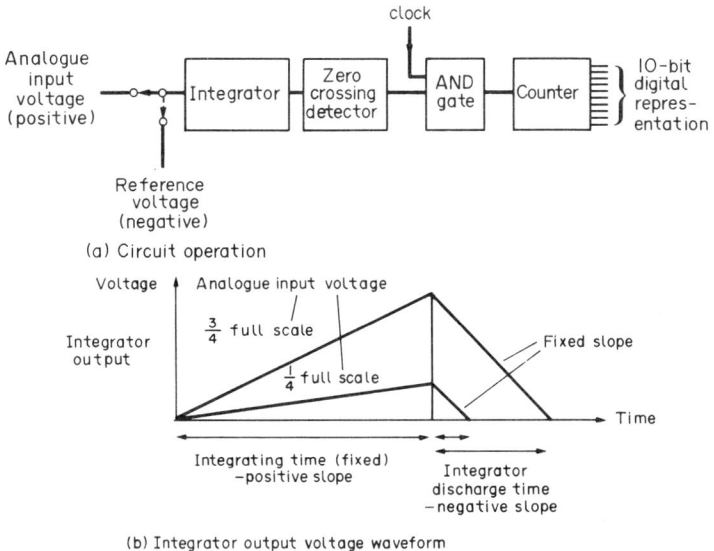

(a) Circuit operation

(b) Integrator output voltage waveform

FIG. 9.3. Dual ramp A/D conversion process.

This technique is slow—conversion speed is typically 40 ms. However it possesses the advantage of reducing the effect of electrical mains noise if the integration period is the same as the period of a mains cycle (20 ms for 50 Hz mains frequency). One full sinewave cycle of mains voltage noise is self-cancelling on the integrator output voltage.

A practical example of an 8-bit successive approximation A/D converter

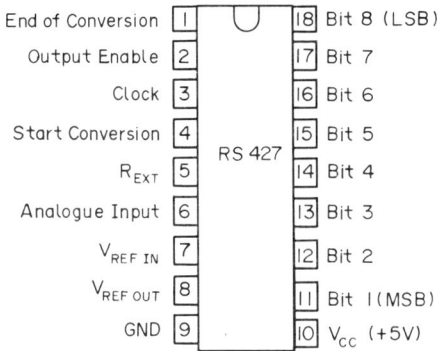

Fig. 9.4. RS 427 A/D converter.

is the RS427, as shown in Fig. 9.4. Commonly the End of Conversion signal is connected to the Output Enable; typical connections for pins 5, 7 and 8 are given in the circuit example in Fig. 9.8, which is described in section 9.3. Clearly the input analogue voltage must not exceed the reference voltage ($V_{REF IN}$).

9.2 DIGITAL TO ANALOGUE (D/A) CONVERSION

A digital to analogue (D/A) converter is used to generate an analogue voltage from a digital electronic system, normally a computer. The analogue voltage may be required to connect to the following:

(a) pen recorder, or chart recorder—to give an historical record of a signal,
(b) graph plotter—to draw shapes,
(c) servo (position control system)—for robot, aerial, gun, etc.,
(d) process control valve—to regulate process flow, temperature, etc.

The pin functions of a D/A IC are simpler than for an A/D, and are shown in Fig. 9.5 for an 8-bit converter. Ten and 12-bit devices are available if greater resolution is required.

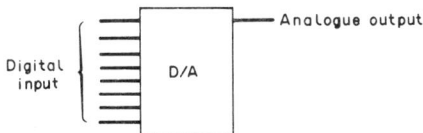

Fig. 9.5. D/A converter pin functions.

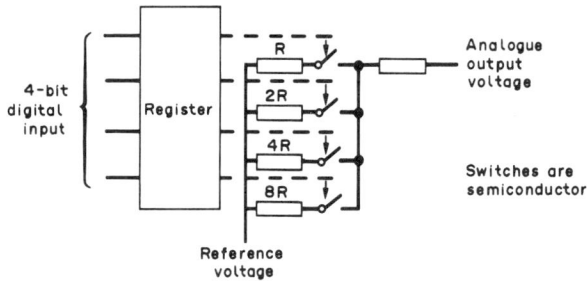

Fig. 9.6. Resistor ladder D/A conversion process.

The most common conversion technique that is applied with D/As is the "resistor ladder" technique, and is illustrated in Fig. 9.6. Only a 4-bit conversion process is illustrated for simplicity. It can be seen that the circuit function is almost identical to the inverse of the successive approximation process for A/D conversion (Fig. 9.2); indeed some devices are offered as either D/A or A/D converters. Each input bit switches a resistor if that bit is set at logic 1. The values of the resistors increase in a binary sequence, so that each switches in a binary weighted fraction ($\frac{1}{2}, \frac{1}{4}, \frac{1}{8}, \frac{1}{16}$) of a reference voltage. Conversion time is very fast—typically 1 μs (1 microsecond).

A practical D/A converter IC is the ZN425E, which is illustrated in Fig. 9.7. The Logic Select is set to logic 0 for D/A conversion, but if set to logic 1 A/D conversion is achieved (if an external comparator circuit is added). The Counter Reset and Clock signals are left unconnected for D/A conversion. This chip is illustrated in the test circuit opposite and described in the following section.

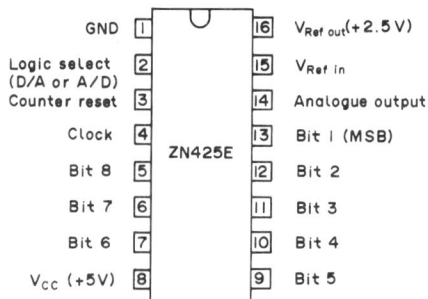

Fig. 9.7. ZN425E D/A converter.

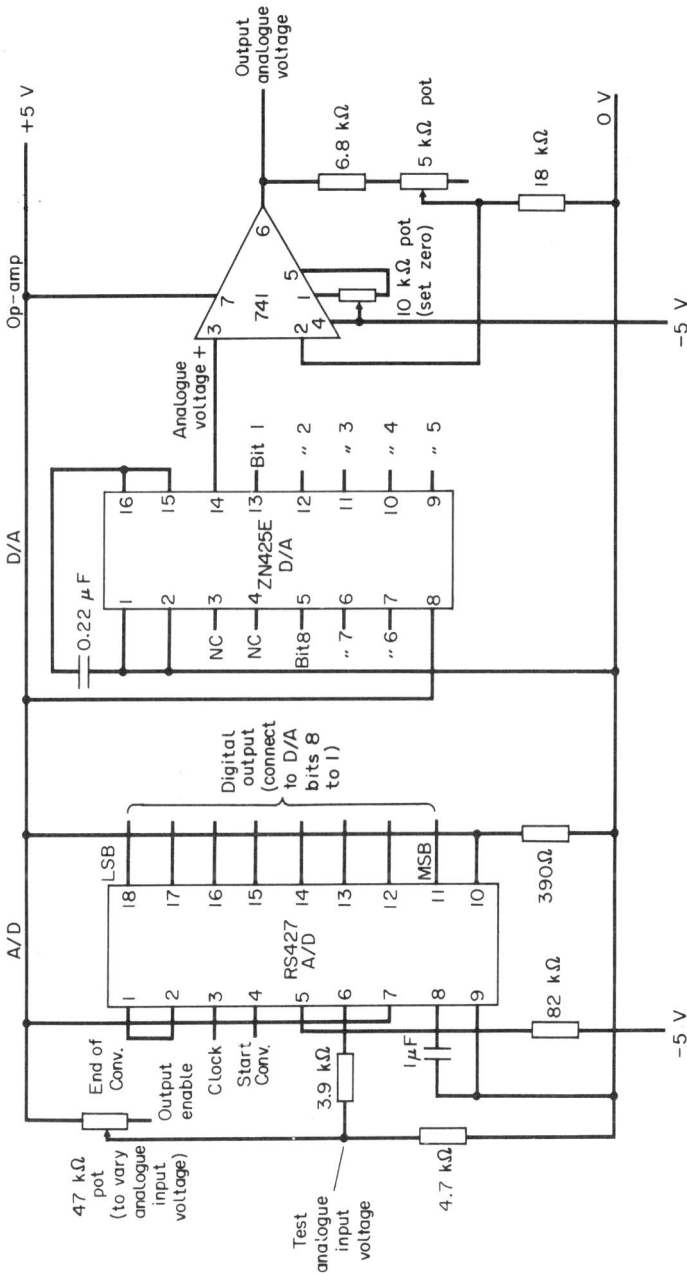

Fig. 9.8. Test A/D and D/A circuit.

9.3 TEST A/D AND D/A CIRCUIT

Normally A/D and D/A ICs connect to computers via 8-bit "ports", as described in chapter 12. A useful test circuit which connects an A/D and a D/A chip in a "back-to-back" test mode, i.e. the digital output of one feeds the input of the other, is illustrated in Fig. 9.8. A variable analogue input voltage is connected to pin 6 on the A/D chip. The digital outputs are permanently enabled—Output Enable is connected to End of Conversion. The A/D circuit can be tested in isolation, and then the digital outputs can be connected to the digital input pins of the D/A. The analogue output voltage of the D/A has a maximum value of +2.5 V, and passes through an op-amp circuit which provides buffering and a calibration mechanism for the overall D/A conversion process.

9.4 ANALOGUE MULTIPLEXING

Frequently a computer requires to read several analogue voltages, e.g. several instrumentation signals. It is often cost-effective to use a single A/D converter chip and to switch each analogue voltage through to this chip in turn. An example of an analogue multiplexor chip is the CD4051BE, which selects one of eight analogue input signals to pass to the single analogue output connection as shown in Fig. 9.9. The particular analogue input signal selected is determined by the binary code on three separate input pins, A, B and C.

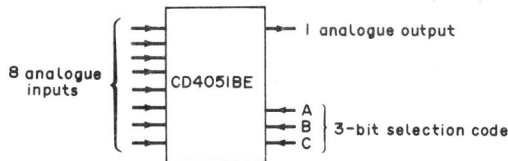

FIG. 9.9. 8 input analogue switch (CD4051BE).

An even simpler alternative is a single IC that combines an analogue multiplexor circuit and an A/D converter within the same package. The 7581 is such a device, and it is illustrated in Fig. 9.10. It accepts eight analogue inputs and using a successive approximation technique sequentially converts each input into an 8-bit digital form. The results of the conversions are stored in eight internal memory locations, each holding eight bits. It is specifically designed to connect to a microcomputer's input port (for the 8-bit digital

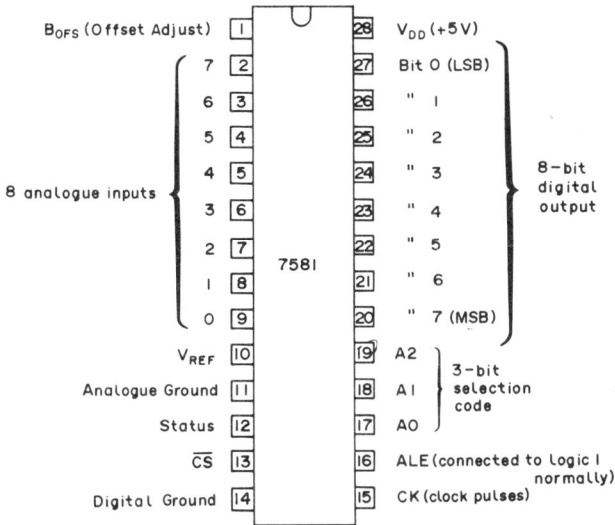

Fig. 9.10. 8 channel, 8-bit, A/D converter (7581).

output) and to an output port (for the 3-bit selection code)—microcomputer ports are described in chapter 12.

9.5 LOGIC LEVEL TO SINEWAVE FREQUENCY CONVERSION

The rapid expansion of computer-based information handling systems in the 1970s and early 1980s, particularly with the advent of the microcomputer, has led to the application of conventional audio sinewave devices for the storage and transmission of binary data. The two most common applications are the use of:

(a) audio cassette recorders for the storage of computer programs and data files—this is particularly common with microcomputers for use in the home;

(b) the telephone network for the transmission of data from computer to remote peripherals/terminals and to other computers.

In both cases there is a requirement to convert binary data into audio frequency sinewaves, and vice versa. Figure 9.11 illustrates these two applications. Each modem in (b) basically consists of the two conversion circuits shown in (a). The technique of using two different frequencies to represent two different logic levels is called FSK (Frequency Shift Keying).

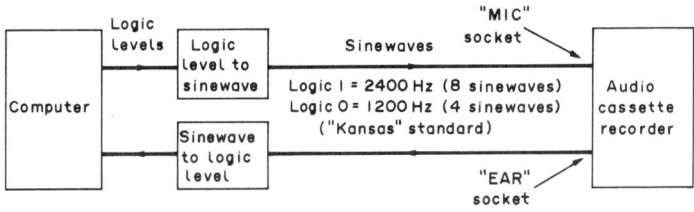

(a) Computer to audio cassette recorder

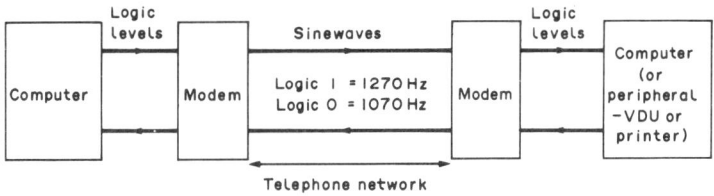

(b) Computer to remote peripheral/computer

FIG. 9.11. Applications of sinewave representation of logic levels.

In this section we will examine a logic level to sinewave frequency conversion circuit. Section 9.6 describes a converter circuit for the reverse function.

Consider Fig. 9.12. Input clock pulses, which can be generated using a 555 timer connected in the astable multivibrator mode (see section 5.1), are divided by two successive D-type flip-flop circuits (see section 4.3). The input logic level signal switches one of these pulse streams (2400 Hz or 1200

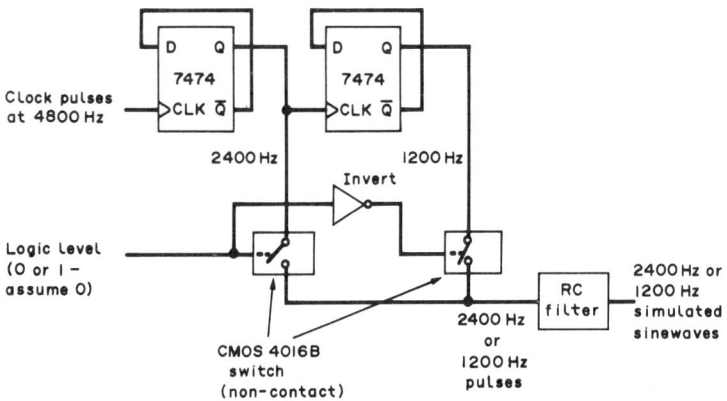

FIG. 9.12. Logic level to sinewave conversion circuit.

Hz) to a simple RC filter circuit which rounds the pulse edges to produce a simulated sinewave output signal.

9.6 SINEWAVE FREQUENCY TO LOGIC LEVEL CONVERSION

An extremely useful IC (the phase locked loop chip) is available to convert a sinewave frequency to a logic level. A typical device is the NE567 and is shown in Fig. 9.13. It can be used to detect one of the two FSK frequencies, so that when this frequency is not detected it can be assumed that the input signal is at the other frequency.

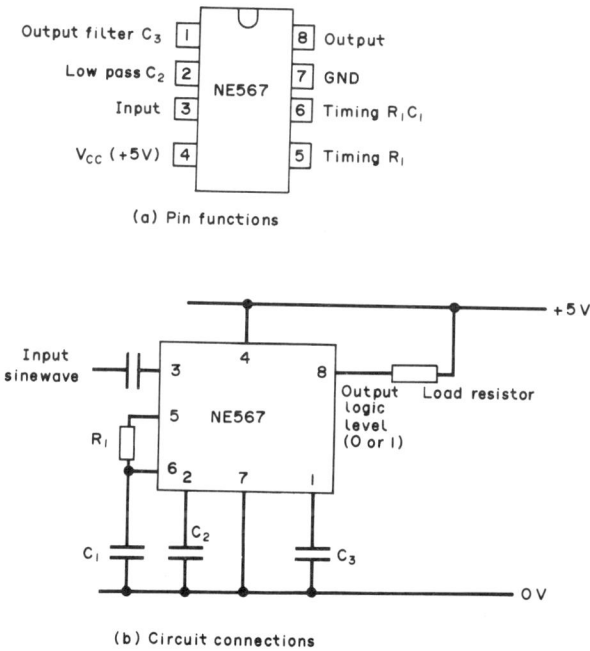

(a) Pin functions

(b) Circuit connections

FIG. 9.13. Phase locked loop (NE567).

Consider a modem application in which it is required to detect a frequency of 1270 Hz. The detection frequency, or "centre frequency", is given by:

$$f_0 = \frac{1.1}{R_1 C_1}$$

and the bandwidth is given by:

ICM-H

$$BW = 1070 \sqrt{\frac{V_1}{f_0 C_2}} \text{ as a } \% \text{ of } f_0$$

where V_1 = input voltage (RMS).
Suitable component values are $R_1 = 1.8\,K\Omega, C_1 = 0.47\,\mu F, C_2 = 10\,\mu F$, C_3 (non-critical) = $10\,\mu F$.

BIBLIOGRAPHY

1. *Microcomputers for Process Control*, R. C. Holland, Pergamon, 1983.
2. *Digital Electronics Practice Using Integrated Circuits*, R. P. Jain and M. M. S. Anand, Tata McGraw-Hill, 1983.
3. *Digital Systems Principles and Applications*, Ronald J. Tocci, Prentice/Hall, 1980.

EXERCISES

1. What is the resolution of a 12-bit A/D converter?
2. The speed of conversion of a successive approximation A/D converter is inversely proportional to the clock rate. Justify this statement in terms of circuit operation of the converter.
3. State an advantage and a disadvantage of a successive approximation A/D converter compared with a dual ramp converter.
4. Sketch the waveform of the analogue input voltage in Fig. 9.3 with an additional electrical mains noise signal superimposed in order to demonstrate the self-cancelling effect that a dual ramp A/D converter exhibits when the integration time = 1/f (where f = mains frequency).
5. Build the circuit of Fig. 9.8 on a "breadboard" and test the circuit in a modular manner—D/A circuit firstly, A/D next, and then integrate the full back-to-back arrangement.
6. Describe an advantage and a disadvantage of multiplexing several instrument analogue signals in a computer plant logging application compared with a system which uses a separate A/D converter for each input signal.
7. Describe the function of a "modem" in data communication systems.
8. Define the "Kansas standard".
9. Design a phase locked loop circuit that "detects" a frequency of 2400 Hz.

CHAPTER 10

Microprocessors

10.1 TYPES OF COMPUTERS

A computer is a programmable data processing system. A computer is digital in operation, and it obeys a list of program instructions which is held in its memory. Each program instruction is simply a group of bits (0s and 1s). Data values which are processed by the program are also held as groups of bits. The number of bits which are used to represent numbers when they are processed (e.g. added) in the machine represents the "word length" of the computer.

There are three basic classifications of computers, as follows:

(a) mainframe computer—large and powerful system which is multi-user and can process large data filing systems (typically 48 or 64 bit word length); used for large company payroll, customer accounts, stores control, etc.

(b) minicomputer—a multi-user system which possesses a smaller number of user-terminals (perhaps up to 20) and processes smaller data filing systems (typically 16 bit word length); used frequently for process monitoring and control applications.

(c) microcomputer—normally a single-user system (most commonly 8 bit word length, although 4 and 16 bit machines are applied) employing a very small number of VLSI ICs; used for a variety of consumer products (pocket calculator, washing machine controller, video game, cash register, etc.) and home and office small computer applications.

10.2 THE BASIC MICROCOMPUTER

Each type of computer can be described diagrammatically using three modules—central processor unit (CPU), memory and input/output. Figure

Address bus (typically 16 lines)

Control bus (typically 12 lines)

Input / output

CPU (Central Processor Unit)

Memory

Data bus (typically 8 lines)

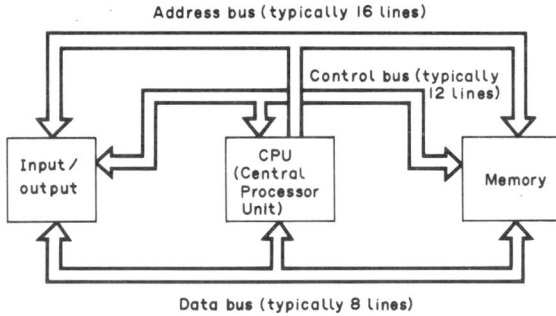

FIG. 10.1. Block diagram of microcomputer.

10.1 shows this diagrammatic representation for a microcomputer. Sometimes these three modules are combined within a single IC, but more commonly the CPU forms a single IC, whilst memory and input/output may each comprise several ICs. When the CPU is a discrete chip it generates the three "buses" shown in order to provide connection to the other ICs.

Memory ICs store words (e.g. 8-bit or "byte" words for an 8-bit CPU) which can represent either of the following:

(a) program instructions which the CPU is to perform, i.e. a list of 8-bit words which the CPU examines and obeys in sequence,
(b) data values which the program of instructions processes, e.g. to perform arithmetic or to transfer through the input/output module.

The normal sequence of actions is that the CPU fetches an instruction using the three buses, examines it and implements it. It then fetches and executes the next instruction, and so on. Occasionally an instruction may demand an input/output data transfer, e.g. send a byte to a VDU (Visual Display Unit) or printer, and the three buses are used to transfer that data value through the input/output module. Typically the input/output module consists of one or more ICs which transfer bytes between CPU and the following:

(1) VDU—possesses CRT (Cathode Ray Tube) for data display and keyboard for data entry,
(2) printer,
(3) floppy disk,
(4) displays, e.g. LEDs, segment display,
(5) plant instrumentation,

and many other devices and circuits.

The roles of the three buses are:

Address bus

To carry the address within memory of the program instruction which is being fetched, or the memory or input/output address of the data value being processed within the instruction.

Data bus

To carry the program instruction, or data value, which has been addressed using the address bus.

Control bus

To carry miscellaneous control signals which are required to activate the above transfers; also special "interrupt" signals which interrupt normal CPU program execution are included.

10.3 THE CPU (MICROPROCESSOR)

The word microprocessor is generally applied to describe the CPU chip; it can also be applied to refer to a chip that provides a CPU function plus some memory and input/output circuits, i.e. a single-chip microcomputer.

Most 8-bit microprocessors are packaged in 40-pin DILs manufactured using MOS technology, and the typical pin functions are listed in Fig. 10.2. Dc power is normally +5 V and 0 V, and the Clock signal is a fast pulse stream of 1 MHz or more which activates each stage in the execution of a single instruction. 16-bit microprocessors are usually packaged in larger chips, e.g. 64-pin, and possess more address lines. It is sensible to concentrate here on 8-bit devices because they are more widely applied still. Also

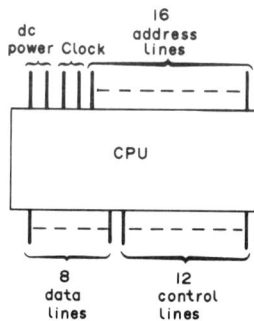

FIG. 10.2. Typical CPU IC pin functions.

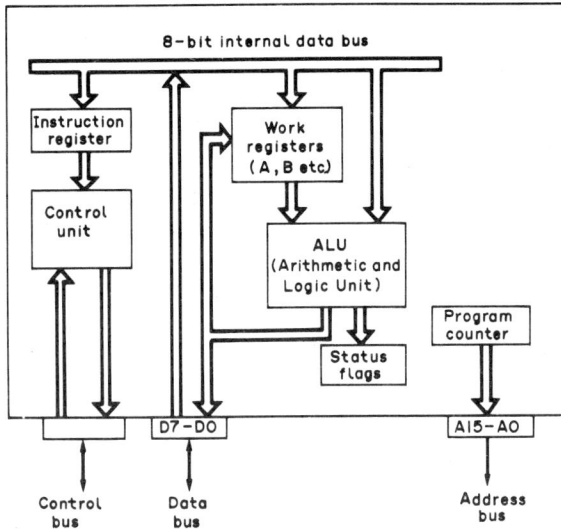

FIG. 10.3. Internal oganisation of CPU.

the principles of CPU operation and interconnection to memory and input/output ICs are simpler to understand with 8-bit devices, but apply also in principle to 16-bit devices.

The simplified internal organisation of a typical 8-bit microprocessor is shown in Fig. 10.3. The roles of the various modules within the chip can be explained by listing the different processes that occur in sequence when an instruction is implemented:

1. *Program counter* contents are sent out on the address bus to locate the next instruction in memory.
2. *Instruction register* receives the instruction byte which is read in from memory along the data bus.
3. *Control unit* examines the instruction byte in the instruction register and implements it by sending out timing and control signals to other parts of the CPU and beyond the CPU.
4. *Work registers* are used to hold data items that are processed within instructions. They are typically between two and seven work registers, and they are commonly called A (accumulator), B, C, etc.
5. *ALU (Arithmetic and Logic Unit)* is used by an instruction whenever arithmetic processing (e.g. add two numbers) or logical operations (e.g. AND two data values) are required. The result of such ALU operations is normally placed in the A register (accumulator).

6. *Status register* (or "flags") is checked occasionally in a program instruction to examine the status of the ALU, e.g. to confirm if the result of the previous instruction passed a zero value out of the ALU. Typically there are five status bits (flags) in the status register to indicate different states of the ALU.

Detailed operation of the CPU when specific instructions are obeyed is described in the following section.

10.4 THE FETCH/EXECUTE CYCLE

The implementation of every instruction by the CPU can be divided into two timing stages:

(a) fetch the instruction (from memory and place in instruction register),
(b) execute the instruction.

Instructions for 8-bit microprocessors are normally 1, 2 or 3 bytes long. Conventionally the first byte, which defines what the instruction is to do, is termed the "opcode", and the second and third bytes are called the "operand". The operand can be a data value (1 byte) that is used in the instruction, or a memory address (2 bytes).

Consider the following three instructions which may form part of a program for a Zilog Z80 microprocessor:

Memory Address	Machine Code	Mnemonic	Comments
3600	C6,04	ADD A,4	;Add 4 to A register (accumulator)
3602	4F	LD C,A	;Transfer contents of A register to C register
3603	3A,20,50	LD A,(5020H)	;Load A register with contents of memory address hexadecimal 5020

The "fetch" operation for each instruction is exactly the same; the contents of the program counter are gated out onto the address bus and the opcode is read in along the data bus and placed in the instruction register.

Figure 10.4 illustrates the operation of the first instruction in this program section. It is assumed that after the CPU had completed the previous instruction the program counter had been incremented automatically to 3600; also it is assumed that the accumulator holds 05 as a result of previous

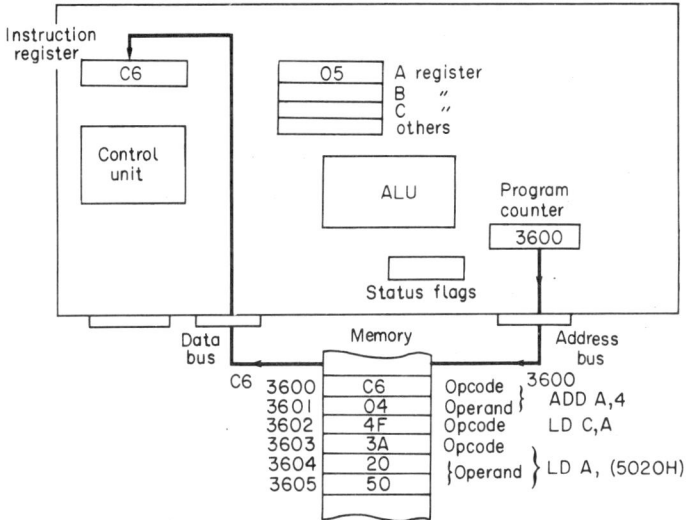

(a) Fetch operation

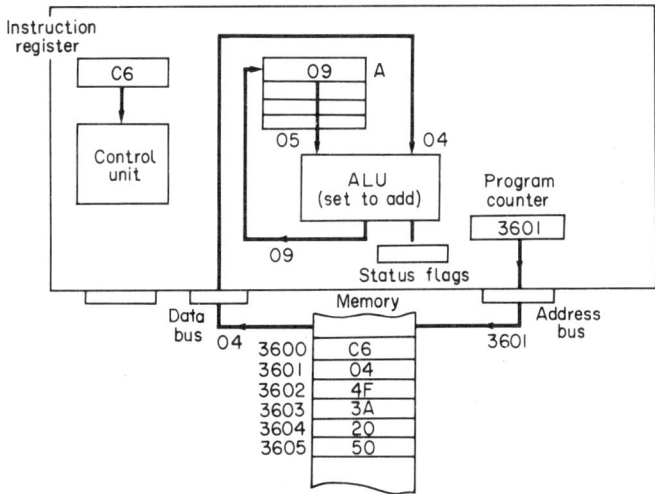

(b) Execute operation

Fig. 10.4. Fetch/execute cycle for ADD A,4 instruction.

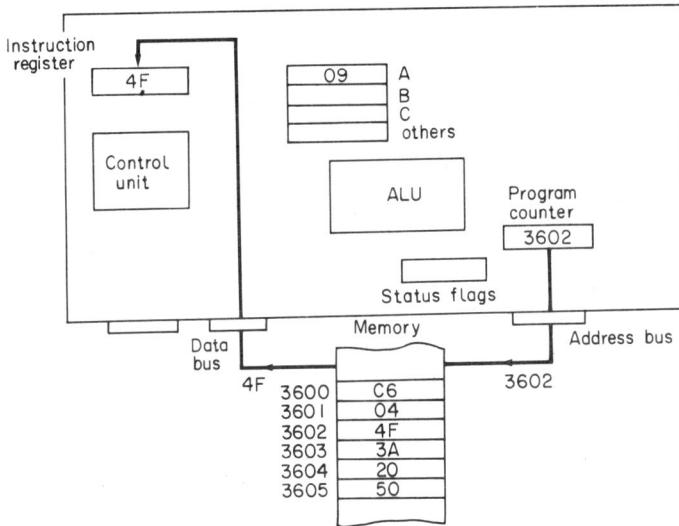

(a) Fetch operation

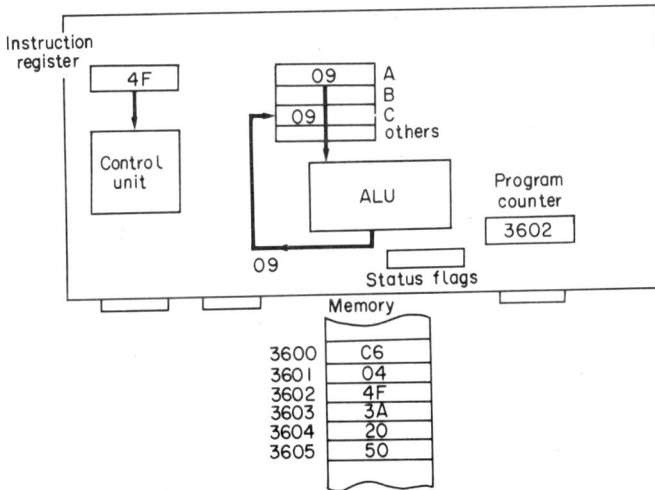

(b) Execute operation

FIG. 10.5. Fetch/execute cycle for LD C,A instruction.

program actions. In the fetch process the memory address 3600 is gated out onto the address bus, and the opcode C6 is read into the CPU along the data bus. It is placed in the instruction register, where it is examined by the control unit during the execute part of the instruction. The execute process in this case demands that the operand 04 (in memory address 3601) is to be read in from memory and added with the contents of the A register (accumulator) 05. The memory address 3601 is sent out along the address bus therefore, and the operand 04 is passed along the data bus and directed to one of the entry points to the ALU. The control unit sets the ALU to perform an add function and releases the two numbers 05 (from A register) and 04 into the ALU simultaneously. It gates the answer 09 back into the A register. The program counter is incremented to 3602 at the end of the instruction to point to the memory address of the following instruction.

Figure 10.5 illustrates the actions performed during the second instruction, which simply moves the contents of A register to C register within the CPU. The execute process does not require a memory transfer of an operand, and consequently this instruction takes less time to implement than the first instruction.

The third instruction requires several stages in the execute procedure, and the stages in its implementation are as follows:

(a) fetch—3A is fetched from memory address 3603;
(b) execute 1—20 is read from memory address 3604 (this is the lower order half of the address of the data value which is to be processed);
(c) execute 2—50 is read from memory address 3605 (higher order half);
(d) execute 3—address 5020 is gated out onto the address bus, and the contents of that memory location are read in along the data bus and placed in the A register.

The program counter is incremented to 3606 when the instruction is completed.

10.5 ZILOG Z80 MICROPROCESSOR

It is generally accepted that the Zilog Z80 is the most powerful 8-bit microprocessor, and it is used in a wide variety of applications as follows:

(a) single board industrial controllers,
(b) home computers, e.g. Sinclair "Spectrum", Tandy "TRS 80",
(c) office computers, e.g. Sharp "MZ80B", Comart "Cromemco C10", Icarus "Superbrain", Research Machines "RML 380Z".

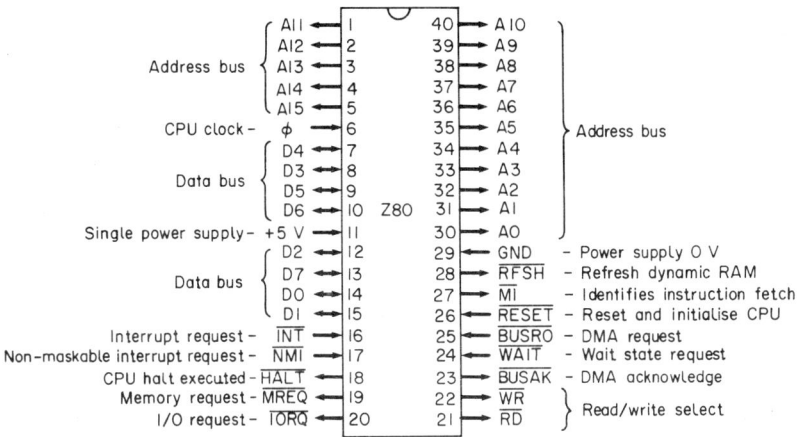

Fig. 10.6. Pin functions of Zilog Z80 microprocessor.

The pin functions of the Z80 are shown in Fig. 10.6. The 40 pins carry the following signals:

(a) 16 address bus lines,
(b) 8 data bus lines,
(c) 2 dc supply lines ($+5$ V and 0 V–GND),
(d) 1 CPU clock pulse line (incoming clock pulses, e.g. 2 MHz),
(e) 13 control bus lines.

In many applications the majority of the control bus lines are not used. The signals that are commonly applied are:

(1) $\overline{MREQ}$ (memory request) and $\overline{IORQ}$ (input/output request)—to select whether memory or input/output is selected by the CPU when a byte is transferred in or out of the CPU;

(2) $\overline{RD}$ (read) and $\overline{WR}$ (write)—to identify the direction of byte transfer;

(3) $\overline{RESET}$—this signal is set when the computer is first switched on to force the CPU to commence obeying program instructions at a fixed memory location (hexadecimal 0000—the start of memory).

Two other control bus lines carry "interrupt" signals—$\overline{INT}$ (interrupt) and $\overline{NMI}$ (non-maskable interrupt). Each of these can be set by an external circuit when it is required to interrupt the program that is running in the microcomputer and implement a special interrupt program. When the interrupt program is completed, return is made to the main program at the point at which the interrupt occurred. When $\overline{NMI}$ is set (to logic 0) an interrupt program at memory address hexadecimal 0066 is entered, and when $\overline{INT}$ is set memory address 0038 is used (although variations of this latter address can be selected). Typical applications of interrupts are with slow peripheral devices, e.g. floppy disks, when it is required to indicate that a required address on the floppy disk surface has been located ready for data transfer. An alternative application is with a regular stream of pulses, e.g. 1 per second, when it is required to maintain a time-of-day as a series of counts in the memory.

Two other control bus signals of interest are $\overline{BUSRQ}$ and $\overline{BUSAK}$, which are used when an input/output IC wishes to access a memory IC directly without data passing through the CPU under program control. This is termed "DMA" (Direct Memory Access). The input/output device requests control of the CPU's three buses by setting $\overline{BUSRQ}$ (to logic 0) and the CPU responds by setting $\overline{BUSAK}$ (to logic 0). The address bus, data bus and some of the control bus lines are then set into the tristate "floating" state (see section 4.6) by the CPU, so that DMA transfers can occur. A typical application of DMA is an input/output circuit that continuously reads data out of the computer's memory in order to generate a video signal to drive a CRT display, e.g. graphics display in a home computer.

The basic diagram of the internal organisation of a CPU shown in Fig. 10.3 applies to the Z80, whilst the block labelled "work registers" for the Z80 is detailed in Fig. 10.7. There are seven main registers (A, B, C, D, E, H and L) which can be used to hold 8-bit data values during program execution. However a second set of registers (A' to L') can be selected (the register sets can be "exchanged")—this is a unique feature with 8-bit microprocessors. Normally this second set of registers is reserved for use by an interrupt program, so that the contents of the registers used in the main program do not have to be stored away in memory when an interrupt program is entered. Notice that the status register (called "flags") is shown as part of this register

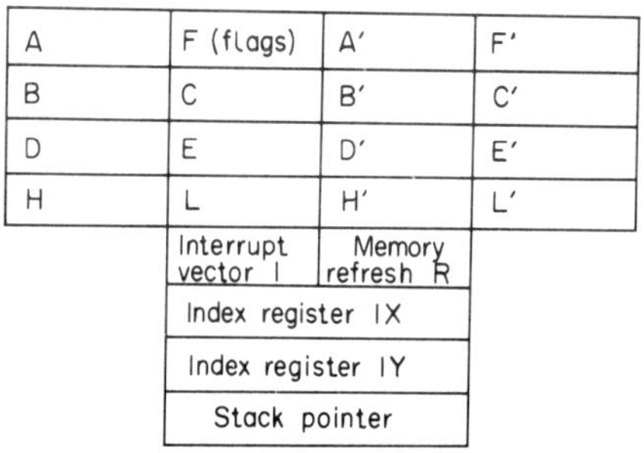

A	F (flags)	A'	F'
B	C	B'	C'
D	E	D'	E'
H	L	H'	L'
	Interrupt vector I	Memory refresh R	
	Index register IX		
	Index register IY		
	Stack pointer		

FIG. 10.7. Work registers for Z80 microprocessor.

block; this is because a new status register is selected when the registers are exchanged. When it is required to access a data item from memory within an instruction, the 16-bit address can be placed in the HL register-pair or in the IX or IY registers. These methods of addressing data items are described in section 10.6.

The interrupt vector is used when it is required to change the start address of an interrupt program. This register contains the top half of the 16-bit memory address, and the interrupting device supplies the bottom half. The CPU must be set into "interrupt mode 2" (using IM 2 instruction) for this method of operation.

The memory refresh register is used to refresh dynamic RAM memory ICs, which are described in chapter 11.

The stack pointer is a 16-bit register which is used to point to a memory address that indicates where in memory the return address (address to which program control must be made when the main program is re-entered) is held when entry is made to an interrupt program or a "subroutine". A subroutine is a section of program that is stored in memory outside the confines of the main program, and it can be called several times by the main program. It is applied to save memory space because the same section of program does not have to be entered several times in the main program. Consider the "memory map" of a program and subroutine shown in Fig. 10.8. When the CALL instruction is obeyed program control transfers to the start of the subroutine and the return address is stored on the stack, which is simply a small block of unused memory. The stack pointer is altered (in this example) from 1FFF to 1FFD. The last instruction in the subroutine is a RET (return) instruction, which causes the return address to be removed from the stack and placed in the CPU's program counter. Program control then resumes in the main program, and the stack pointer returns to 1FFF. Several CALL instructions to the same subroutine can be made within the main program. If a subroutine calls another subroutine, the second return address is stored on top of the first return address in the stack. The stack pointer always contains the memory address of the last *used* location in the stack.

Although the primary function of the stack is to store return addresses, it can also be used to store the contents of the work registers when a subroutine (or interrupt program) is entered. This is necessary to prevent the subroutine overwriting the registers which are used by the main program. A PUSH instruction is used to store a register (or normally two registers) on the stack, and a POP instruction is used to remove data values off the stack and reinstate them in registers. These instructions, which are listed in the following section, are normally used at the start and end of a subroutine respectively.

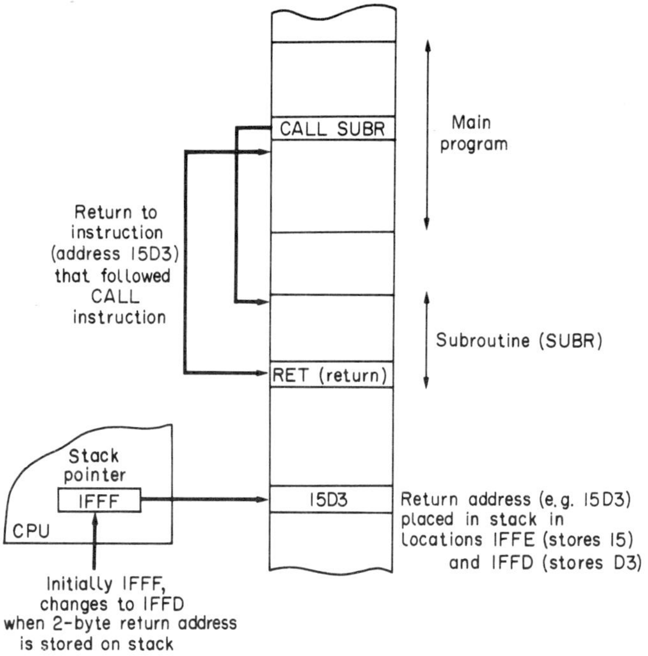

FIG. 10.8. Storage of return address on stack.

10.6 ZILOG Z80 INSTRUCTION SET

The instruction set for the Zilog Z80 microprocessor is listed in Table 10.1 on pp. 120–123. In order to follow the operation of many of these instructions it is necessary to understand the various "addressing modes", i.e. methods of addressing data values which are to be used within instructions. Examples of the five Z80 addressing modes are:

(a) *Direct register*, e.g.

    LD   A,B   ;Load A register with contents of B register

"Destination"   "Source"

In this case the destination and source both use direct register addressing, i.e. the data item is held initially in a CPU work register (B) and is transferred to another work register (A). It is also retained in B.

(b) *Direct memory*, e.g.

    LD   A,(1900H)   ;Load A register with contents of memory
                           ;location hexadecimal 1900

In this case the source is a memory location. The instruction is 3 bytes long; the operand 1900 fills the second and third bytes. The destination is of course direct register, as for all these examples.

(c) *Indirect register*, e.g.

    LD   A,(HL)   ;Load A register with the contents of memory
                       ;location which is held in HL register-pair

In this case the HL register-pair "point to" the memory location that contains the data value. The HL register-pair must be loaded with the required 16-bit memory address by a previous instruction in the program. The H and L registers are normally reserved within a program for this type of addressing.

(d) *Immediate*, e.g.

    LD   A,19   ;Load A register with decimal 19

In this case the data value is not held in a register within the CPU or in a memory location. It is contained within the instruction itself—it forms the second byte of the instruction.

(e) *Indexed*, e.g.

    LD   A,(IX+8)   ;Load A register with the contents of memory
                         ;location which is held in IX register, with 8
                         ;added to that memory location

In this case assume that IX had been loaded previously with memory address hexadecimal 4000. The instruction loads A register with the contents of memory location 4008. The "displacement" d (8 in this case) forms the fourth byte of a 4-byte instruction. If d is zero, the instruction is simply indirect register addressing as described in (c) above (using IX in place of HL).

There are two methods of referencing memory addresses when "jump" instructions are inserted into a program. Jump instructions represent a discontinuity in the program flow and cause program execution to transfer to a different area of memory. The two methods are:

(f) *Absolute jump*, e.g.

    JP   183AH   ;Jump to memory address hexadecimal 183A

The following instruction (if one exists) is not obeyed. Program control transfers to memory address 183A. This address 183A forms the second and third bytes of the instruction.

(g) *Relative jump*, e.g.

    JR   5   ;Jump 5 memory locations (from the address of the
                    ;following instruction)

Normally when an instruction is completed the program counter is set to the address of the following instruction. In this case 5 is added to that address. The displacement (5 in this case) is a two's complement number and can take any value in the range -127 to $+127$. It forms the second byte of the instruction.

These two jump instructions are "unconditional" jumps, i.e. they must be obeyed. "Conditional" jumps are only obeyed if a bit in the status register is set, e.g.

    JP   NZ,1807H   ;Jump if non zero (if the zero status bit is not
                         ;set by the previous instruction) to memory
                         ;address 1807
    JR   Z,-8 ;Jump if zero (if the zero status bit is set by the
                    ;previous instruction) back 8 memory locations

It is often convenient to group instructions within the instruction set into four categories, as follows:

(1) *Data move instructions*, e.g.

    LD   D,B   ;Load D register with contents of B register
    LD   (18A9H),C   ;Load memory location 18A9 with contents of C
                         ;register
    OUT (01H),A   ;Output contents of A register to input/output
                         ;address 01

(2) *Data modify instructions*, e.g.

    ADD A,3   ;Add 3 to A register
    DEC B   ;Decrement (subtract 1 from) B register
    AND 0FH   ;Logical AND binary 0000 1111 with A register
    RL   B   ;Rotate B register left 1 bit position

(3) *Jump instructions*, e.g.

    as above in (f) and (g).

(4) *Miscellaneous control instructions*, e.g.

| HALT | ;Stop program execution |
| EI | ;Enable interrupts |

The use of this instruction set is best illustrated by a few sample programs:

Example Program 1
Examine the contents of memory location 0700, add 6 to it, and store the answer in memory location 0800.

LD	A,(0700H)	;Load A register with contents of memory ;location 0700
ADD	A,6	;Add A,6 to A register
LD	(0800H),A	;Store A register in memory location 0800
HALT		;Stop program

Example program 2
Examine the contents of memory location 2000 and overwrite the contents with 0 if they are FF initially.

LD	A,(2000H)	;Load A register with contents of memory ;location 2000
CP	FFH	;Compare A register with FF
JP	NZ,FINISH	;Jump if not zero (data values are different)
LD	A,0	;Load A register with 0
LD	(2000H),A	;Store A register in memory location 2000
	FINISH:HALT	;Stop program

Example Program 3
Add the contents of an 8-value data list held in memory starting at 5000, store the answer in memory location 6000 and then enter another program that starts at memory location 0000.

LD	A,0	;Clear A register
LD	HL,5000H	;Load HL register-pair with 5000
LD	B,8	;Set loop count of 8 in B register
	LOOP:ADD A,(HL)	;Add value from data list into A register
INC	HL	;Increment HL (select next data value)
DEC	B	;Decrement loop count
JR	NZ,LOOP	;Repeat loop until loop count is zero
LD	(6000H),A	;Store answer in memory location 6000
JP	0000H	;Jump to other program at memory location

TABLE 10.1   ZILOG Z80 INSTRUCTION SET

Mnemonic	Description	Machine code
ADC HL,ss	Add with carry reg. pair ss to HL	ED,4A (ss=BC)
ADC A,s	Add with carry operand s to Acc.	CE,20 (s=20)
ADD A,n	Add value n to Acc.	C6,20 (n=20)
ADD A,r	Add reg. r to Acc.	80 (r=B)
ADD A,(HL)	Add location (HL) to Acc.	86
ADD A,(IX+d)	Add location (IX+d) to Acc.	DD,86,05 (d=05)
ADD A,(IY+d)	Add location (IY+d) to Acc.	FD,86,05 (d=05)
ADD HL,ss	Add reg. pair ss to HL	09 (ss=BC)
ADD IX,pp	Add reg. pair pp to IX	DD,09 (pp=BC)
ADD IY,rr	Add reg. pair rr to IY	FD,09 (rr=BC)
AND s	Logical AND of operand s and Acc.	E6,20 (s=20)
BIT b,(HL)	Test BIT b of location (HL)	CB,46 (b=0)
BIT b,(IX+d)	Test BIT b of location (IX+d)	DD,CB,05,46 (b=0, d=05)
BIT b,(IY+d)	Test BIT b of location (IY+d)	FD,CB,05,46 (b=0,d=05)
BIT b,r	Test BIT b of reg. r	CB,47 (b=0,r=A)
CALL cc,nn	Call subroutine at location nn if condition cc is true	DC,84,05 (cc=C,nn=0584)
CALL nn	Unconditional call subroutine at location nn	CD,84,05 (nn=0584)
CCF	Complement carry flag	3F
CP s	Compare operand s with Acc.	FE,20 (s=20)
CPD	Compare location (HL) and Acc., decrement HL and BC	ED,A9
CPDR	Compare location (HL) and Acc., decrement HL and BC, repeat until BC=0	ED,B9
CPI	Compare location (HL) and Acc., increment HL and decrement BC	ED,A1
CPIR	Compare location (HL) and Acc., increment HL, decrement BC, repeat until BC=0	ED,B1
CPL	Complement Acc. (1s comp.)	2F
DAA	Decimal adjust Acc.	27
DEC m	Decrement operand m	3D (m=A) 05 (m=B)
DEC (HL)	Decrement (HL)	35
DEC IX	Decrement IX	DD,2B
DEC IY	Decrement IY	FD,2B
DEC ss	Decrement reg. pair ss	0B (ss=BC)
DI	Disable interrupts	F3
DJNZ e	Decrement B and jump relative if B≠0	10,2E (e=2E)
EI	Enable interrupts	FB
EX (SP),HL	Exchange the location (SP) and HL	E3
EX (SP),IX	Exchange the location (SP) and IX	DD,E3
EX (SP),IY	Exchange the location (SP) and IY	FD,E3
EX AF,AF′	Exchange the content of AF and AF′	08
EX DE,HL	Exchange the contents of DE and HL	EB
EXX	Exchange the contents of BC,DE,HL with contents of BC′,DE′,HL′	D9

TABLE 10.1   (contd.)

Mnemonic	Description	Machine code
HALT	Halt (wait for interrupt or reset)	76
IM 0	Set interrupt mode 0	ED,46
IM 1	Set interrupt mode 1	ED,56
IM 2	Set interrupt mode 2	ED,5E
IN A,(n)	Load Acc. with input from device n	DB,20 (n=20)
IN r,(C)	Load reg. r with input from device (C)	ED,78 (r=A)
INC (HL)	Increment location (HL)	34
INC IX	Increment IX	DD,23
INC (IX+d)	Increment location (IX+d)	DD,34,05 (d=05)
INC IY	Increment IY	FD,23
INC (IY+d)	Increment location (IY+d)	FD,34,05 (d=05)
INC r	Increment reg. r	3C (r=A)
		04 (r=B)
INC ss	Increment reg. pair as	03 (ss=BC)
IND	Load location (HL) with input from port (C), decrement HL and B	ED,AA
INDR	Load location (HL) with input from port (C), decrement HL and decrement B, repeat until B=0	ED,BA
INI	Load location (HL) with input from port (C), increment HL and decrement B	ED,A2
INIR	Load location (HL) with input from port (C), increment HL and decrement B, repeat until B=0	ED,B2
JP (HL)	Unconditional jump to (HL)	E9
JP (IX)	Unconditional jump to (IX)	DD,E9
JP (IY)	Unconditional jump to (IY)	FD,E9
JP cc,nn	Jump to location nn if condition cc is true	DA,84,05 (cc=C,nn=0584)
JP nn	Unconditional jump to location nn	C3,84,05 (nn=0584)
JR C,e	Jump relative to PC+e if carry=1	38,2E (e=2E)
JR e	Unconditional jump relative to PC+e	18,2E (e=2E)
JR NC,e	Jump relative to PC+e if carry=0	30,2E (e=2E)
JR NZ,e	Jump relative to PC+e if non zero (Z=0)	20,2E (e=2E)
JR Z,e	Jump relative to PC+e if zero (Z=1)	28,2E (e=2E)
LD A,(BC)	Load Acc. with location (BC)	0A
LD A,(DE)	Load Acc. with location (DE)	1A
LD A,I	Load Acc. with I	ED,57
LD A,(nn)	Load Acc. with location nn	3A,84,05 (nn=0584)
LD A,r	Load Acc. with reg. r	78 (r=B)
LD (BC),A	Load location (BC) with Acc.	02
LD (DE),A	Load location (DE) with Acc.	12
LD (HL),n	Load location (HL) with value n	36,20 (n=20)
LD dd,nn	Load reg. pair dd with value nn	01,84,05 (dd=BC,nn=0584)
LD dd,(nn)	Load reg. pair dd with location (nn)	ED,4B,84,05 (dd=BC,nn=0584)
LD HL,(nn)	Load HL with location (nn)	2A,84,05 (nn=0584)
LD HL,nn	Load HL with nn	21,84,05 (nn=0584)

(*continued*)

TABLE 10.1 (contd.)

Mnemonic	Description	Machine code
LD (HL),r	Load location (HL) with reg. r	77 (r=A)
LD I,A	Load I with Acc.	ED,47
LD IX,nn	Load IX with value nn	DD,21,84,05 (nn=0584)
LD IX,(nn)	Load IX with location (nn)	DD,2A,84,05 (nn=0584)
LD (IX+d),n	Load location (IX+d) with value n	DD,36,05,20 (n=20,d=05)
LD (IX+d),r	Load location (IX+d) with reg. r	DD,77,05 (r=A,d=05)
LD IY,nn	Load IY with value nn	FD,21,84,05 (nn=0584)
LD IY,(nn)	Load IY with location (nn)	FD,2A,84,05 (nn=0584)
LD (IY+d),n	Load location (IY+d) with value n	FD,36,05,20 (n=20,d=05)
LD (IY+d),r	Load location (IY+d) with reg. r	FD,77,05 (r=A,d=05)
LD (nn),A	Load location (nn) with Acc.	32,84,05 (nn=0584)
LD (nn),dd	Load location (nn) with reg. pair dd	ED,43,84,05 (dd=BC,nn=0584)
LD (nn),HL	Load location (nn) with HL	22,84,05 (nn=0584)
LD (nn),IX	Load location (nn) with IX	DD,22,84,05 (nn=0584)
LD (nn),IY	Load location (nn) with IY	FD,22,84,05 (nn=0584)
LD R,A	Load R with Acc.	ED,4F
LD r,(HL)	Load reg. r with location (HL)	7E (r=A)
LD r,(IX+d)	Load reg. r with location (IX+d)	DD,7E,05 (r=A,d=05)
LD r,(IY+d)	Load reg. r with location (IY+d)	FD,7E,05 (r=A,d=05)
LD r,n	Load reg. r with value n	3E,20 (r=A,n=20)
LD SP,HL	Load SP with HL	F9
LD SP,IX	Load SP with IX	DD,F9
LD SP,IY	Load SP with IY	FD,F9
LDD	Load location (DE) with location (HL), decrement DE, HL and BC	ED,A8
LDDR	Load location (DE) with location (HL), decrement DE, HL and BC, repeat until BC=0	ED,B8
LDI	Load location (DE) with location (HL), increment DE, HL, decrement BC	ED,A0
LDIR	Load location (DE) with location (HL), increment DE, HL, decrement BC and repeat until BC=0	ED,B0
NEG	Negate Acc. (2s complement)	ED,44
NOP	No operation	00
OR s	Logical OR of operand s and Acc.	F6,20 (s=20)
OTDR	Load output port (C) with location (HL), decrement HL and B, repeat until B=0	ED,8B
OTIR	Load output port (C) with location (HL), increment HL, decrement B, repeat until B=0	ED,B3
OUT (C),r	Load output port (C) with reg. r	ED,79 (r=A)
OUT (n),A	Load output port (n) with Acc.	D3,20 (n=20)
OUTD	Load output port (C) with location (HL), decrement HL and B	ED,AB
OUTI	Load output port (C) with location (HL), increment HL and decrement B	ED,A3
POP IX	Load IX with top of stack	DD,E1

TABLE 10.1 (contd.)

Mnemonic	Description	Machine code
POP IY	Load IY with top of stack	FD,E1
POP qq	Load reg. pair qq with top of stack	F1 (qq=AF)
PUSH IX	Load IX onto stack	DD,E5
PUSH IY	Load IY onto stack	FD,E5
PUSH qq	Load reg. pair qq onto stack	F5 (qq=AF)
RES b,m	Reset bit b of operand m	CB,87 (b=0,m=A)
RET	Return from subroutine	C9
RET cc	Return from subroutine if condition cc is true	D8 (cc=C)
RETI	Return from interrupt	ED,4D
RETN	Return from non-maskable interrupt	ED,45
RL m	Rotate left through carry operand m	CB,17 (m=A)
RLA	Rotate left Acc. through carry	17
RLC (HL)	Rotate location (HL) left circular	CB,06
RLC (IX+d)	Rotate location (IX+d) left circular	DD,CB,05,06 (d=05)
RLC (IY+d)	Rotate location (IY+d) left circular	FD,CB,05,06 (d=05)
RLC r	Rotate reg. r left circular	CB,07 (r=A)
RLCA	Rotate left circular Acc.	07
RLD	Rotate digit left and right between Acc. and location (HL)	ED,6F
RR m	Rotate right through carry operand m	CB,1F (m=A)
RRA	Rotate right Acc. through carry	1F
RRC m	Rotate operand m right circular	CB,0F (m=A)
RRCA	Rotate right circular Acc.	0F
RRD	Rotate digit right and left between Acc. and location (HL)	ED,67
RST p	Restart to location p	C7 (p=00H)
SBC A,s	Subtract operand s from Acc. with carry	98 (s=B)
SBC HL,ss	Subtract reg. pair ss from HL with carry	ED,42 (ss=BC)
SCF	Set carry flag (C=1)	37
SET b,(HL)	Set bit b of location (HL)	CB,C6 (b=0)
SET b,(IX+d)	Set bit b of location (IX+d)	DD,CB,05,C6 (b=0,d=05)
SET b,(IY+d)	Set bit b of location (IY+d)	FD,CB,05,C6 (b=0, d=05)
SET b,r	Set bit b of reg. r	CB,C7 (b=0,r=A)
SLA m	Shift operand m left arithmetic	CB,27 (m=A)
SRA m	Shift operand m right arithmetic	CB,2F (m=A)
SRL m	Shift operand m right logical	CB,3F (m=A)
SUB s	Subtract operand s from Acc.	90 (s=B)
XOR s	Exclusive OR operand s and Acc.	AB (s=B)

10.7 INTEL 8085 MICROPROCESSOR

The Intel 8085 microprocessor is very similar to the Z80. Machine code programs written for the 8085 will run on the Z80. However reverse compatibility is not total because the Z80 possesses several extra instructions compared with the 8085.

```
      Crystal   X1 ──→ │ 1        40 │ ◄── V_cc (+5 V) - Power supply +5 V
     (or RC)   X2 ──→ │ 2        39 │ ◄── HOLD  } DMA
Peripherals reset-RESET OUT ◄── │ 3   38 │ ──► HLDA
Serial output and input { SOD ◄── │ 4  37 │ ──► CLOCK (OUT) - Clock for peripherals
                         SID ──→ │ 5  36 │ ◄── RESET IN - System reset
               TRAP ──→ │ 6       35 │ ◄── READY - Wait state request
             RST 7.5 ──→ │ 7      34 │ ──► IO/M̄ - I/O or memory select
  Interrupts { RST 6.5 ──→ │ 8    33 │ ──► SI - Bus state indicator
             RST 5.5 ──→ │ 9      32 │ ──► RD } Read/write select
               INTR ──→ │ 10 8085 31 │ ──► WR
INTR acknowledge - INTRA ◄── │ 11  30 │ ──► ALE - Address latch enable
             ┌ AD0 ◄──► │ 12     29 │ ──► SO - Bus state indicator
             │ AD1 ◄──► │ 13     28 │ ──► A15 ┐
             │ AD2 ◄──► │ 14     27 │ ──► A14 │
 Multiplexed │ AD3 ◄──► │ 15     26 │ ──► A13 │
 address and │ AD4 ◄──► │ 16     25 │ ──► A12 │ Address bus
 data bus    │ AD5 ◄──► │ 17     24 │ ──► A11 │
             │ AD6 ◄──► │ 18     23 │ ──► A10 │
             └ AD7 ◄──► │ 19     22 │ ──► A9  │
Power supply - V_ss (OV) ──→ │ 20  21 │ ──► A8 ┘
         OV
```

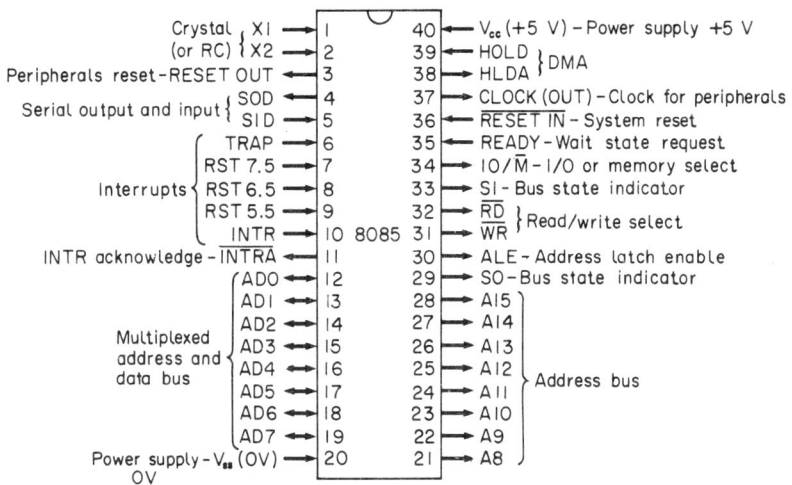

Fig. 10.9. Pin functions of Intel 8085 microprocessor.

The 8085 was the first widely applied microprocessor and has been used extensively in factory monitoring and control applications. It is not used in any popular home or office microcomputers.

The pin functions of the 8085 are illustrated in Fig. 10.9. The most significant difference between the 8085 and the Z80 is that the data bus with the Intel device is multiplexed with the lower half of the address bus, i.e. both sets of signals share the same pins (AD0 to AD7). Normally an additional IC is required with the 8085 to de-multiplex these signals. The ALE signal identifies whether address (ALE = logic 1) or data (ALE = logic 0) is set on the multiplexed lines.

The device possess an on-chip CPU clock circuit, and so only an external crystal must be connected to the clock pins 1 and 2.

The 8085 possesses an unusual feature, which is not offered by other microprocessors, and that is the inclusion of single-bit output (SOD) and input (SID) signals.

The work registers for the 8085 are shown in Fig. 10.10. This is a much smaller register set compared with the Z80. There is only one main register block (A, B, C, D, E, H and L) and there are no index registers.

Notice that although the machine code for all Intel 8085 instructions is the same as for the Zilog Z80 instructions, different mnemonics are used. Therefore a program version written in mnemonic form ("assembly language") for the 8085 appears different to a Z80 program, although the machine code version that is held in memory during program execution is identical.

A	
B	C
D	E
H	L
Stack pointer	

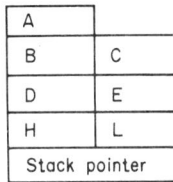

Fig. 10.10. Work registers for 8085 microprocessor.

10.8 MOS TECHNOLOGY 6502 MICROPROCESSOR

The manufacturer MOS Technology produces one of the most popular microprocessors that is used in home computers. The 6502 is applied in the BBC, Vic 20, Oric, Commodore PET and Apple microcomputers. It is also used in a proprietary feedback controller for a continuous process in a factory application (called "three-term controller").

Figure 10.11 shows the pin functions of the 6502. They are similar to those

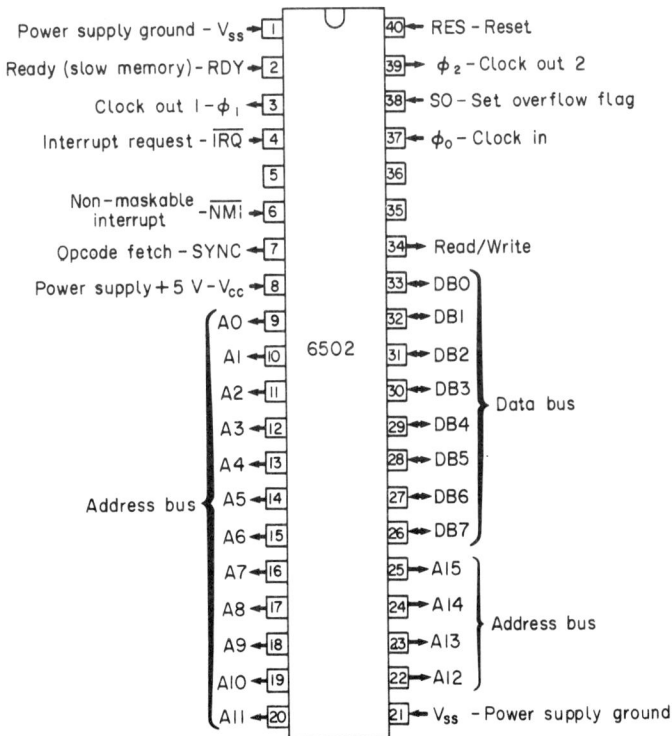

Power supply ground – V_{ss} – [1]   [40] – RES – Reset
Ready (slow memory) – RDY – [2]   [39] – ϕ_2 – Clock out 2
Clock out I – ϕ_1 – [3]   [38] – SO – Set overflow flag
Interrupt request – $\overline{IRQ}$ – [4]   [37] – ϕ_0 – Clock in
[5]   [36]
Non–maskable interrupt – $\overline{NMI}$ – [6]   [35]
Opcode fetch – SYNC – [7]   [34] – Read/Write
Power supply + 5 V – V_{cc} – [8]   [33] – DB0
A0 – [9]   [32] – DB1
A1 – [10]   [31] – DB2
A2 – [11]   [30] – DB3
A3 – [12]   [29] – DB4   Data bus
A4 – [13]   [28] – DB5
A5 – [14]   [27] – DB6
A6 – [15]   [26] – DB7
A7 – [16]   [25] – A15
A8 – [17]   [24] – A14
A9 – [18]   [23] – A13   Address bus
A10 – [19]   [22] – A12
A11 – [20]   [21] – V_{ss} – Power supply ground

Address bus { A0–A11 }   6502

Fig. 10.11. Pin functions of MOS Technology 6502 microprocessor.

| Accumulator A |
| Index register X |
| Index register Y |
| Stack pointer |

All 8-bit registers

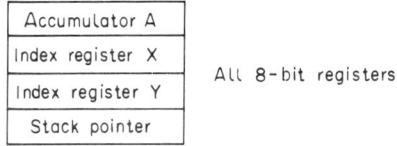

FIG. 10.12. Work registers for 6502 microprocessor.

for the Zilog Z80 (see Fig. 10.6). The SYNC signal performs the same function as M1 with the Z80. Similarly RDY is synonymous with the Z80's WAIT.

The work registers for the 6502 are illustrated in Fig. 10.12. This is a much more limited register set compared with the Z80 and the 8085. There are only three registers (A, X and Y) that can be used to hold data values during program operation. Also the stack pointer is only 8-bits wide; therefore the stack is limited to a small area of memory (address decimal 256 to 511).

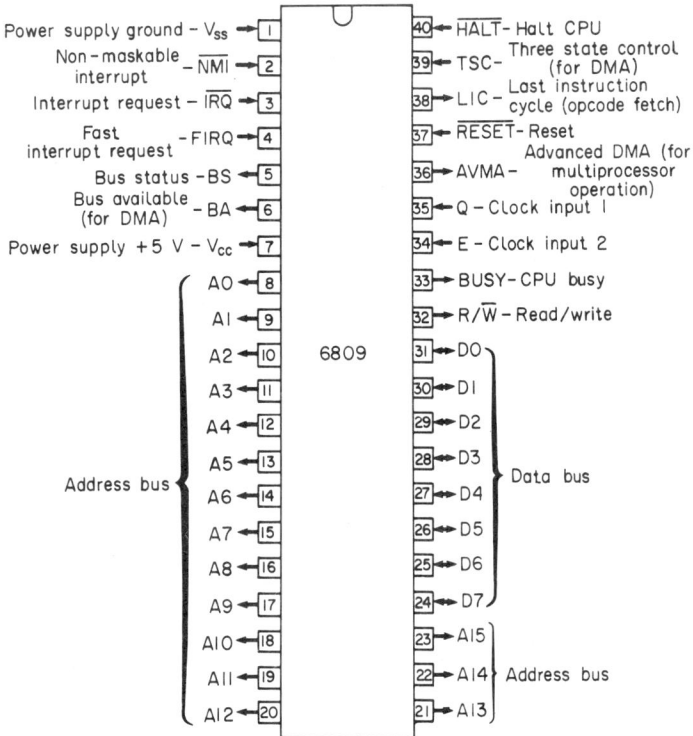

FIG. 10.13. Pin functions of Motorola 6809 microprocessor.

10.9 MOTOROLA 6809 MICROPROCESSOR

Motorola introduced the 6809 as an update on the earlier 6800. The 6809 is applied in the Dragon 32 home computer, but despite an attractive range of features it has not achieved such widespread application as the other 8-bit microprocessors described previously.

The pin functions are detailed in Fig. 10.13. Notice that there are three interrupt signals in addition to Reset. The usual DMA control signals are available.

Figure 10.14 details the 6809 work registers. Both registers A and B are accumulators, i.e. the results of ALU operations, e.g. arithmetic functions, can be placed in either register. Two 16-bit indexing registers are available for indexed addressing mode operation. The CPU uses the hardware stack pointer during subroutine calls and interrupts. The programmer controls the user stack pointer and can pass data values to subroutines using this stack.

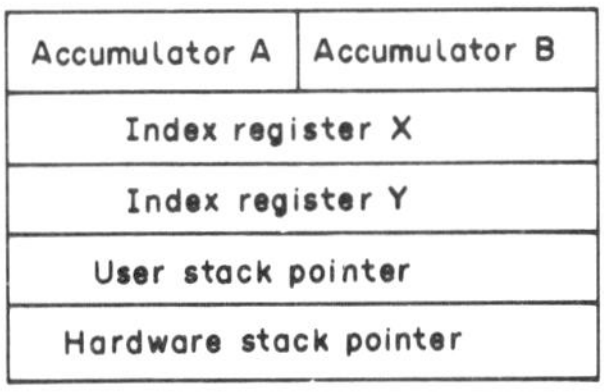

Accumulator A	Accumulator B
Index register X	
Index register Y	
User stack pointer	
Hardware stack pointer	

Fig. 10.14. Work registers for 6809 microprocessor.

10.10 16-BIT MICROPROCESSORS

All of the microcomputers described so far are 8-bit devices. 16-bit microcomputers offer several advantages, e.g.

(a) larger number range—64 K compared with 256
(b) more work registers
(c) more addressing modes
(d) more instructions, e.g. multiply and divide (not offered by any 8-bit devices)
(e) wider addressing range, e.g. 1 M byte in place of 64 K byte
(f) increased speed of operation

Despite these advantages they have not replaced 8-bit devices on a widespread scale due to the increased chip cost (typically 30 times more expensive) and the investment (particularly in supporting software) that has been generated for the earlier 8-bit components. In many single-user applications, e.g. home computer and office computer, an 8-bit micro-computer is adequate, despite slow speed and smaller memory capacity. Similarly for a small application, e.g. petrol pump controller or industrial controller, an 8-bit device performs its task adequately. However 16-bit devices have been introduced in many office computers, particularly when multi-user in operation.

The 8-bit memory and input/output chips (described in the next two chapters) that have been designed to support 8-bit microprocessors are used also with 16-bit CPUs; normally they are used in pairs. There are no 16-bit support chips.

The main 16-bit microprocessors are:

(1) *Intel 8086 and 8088*

The 8086 possesses 12 work registers, 8 of which are accumulators, 20 address bus lines (giving 1 M word addressing range) and extremely fast instruction execution times (e.g. 250 ns add time). It is widely used in the new British telephone switching system ("System X"). The 8088 is virtually identical to the 8086, except that its external data bus is only 8 bits wide in place of 16 bits. It has become almost a standard for 16-bit office computers, e.g. IBM Personal Computer, ACT Sirius and DEC Rainbow.

(2) *Motorola 68000*

The 68000 is arguably the most powerful 16-bit microprocessor. It possesses 15 work registers, each 32 bits wide, and it has often been claimed to be a 32-bit device, although its internal data bus is only 16 bits wide. It possesses 23 address lines and is packages in a 64-pin package. It is used in multi-user microcomputer systems, when it rivals minicomputers in terms of speed and computing power, and in industrial logging applications.

(3) *Texas Instruments 9900*

The Texas Instruments 9900 series of 16-bit microprocessors was the first 16-bit range of devices to be offered commercially. The processors are very slow in operation, but offer some unusual and advantageous features. In particular there are no work registers within the CPU chip—they are held in memory. This enables an interrupt program to switch to another set of registers within memory, without the necessity of storing away the main program's registers contents. Therefore it does not use a stack. 9900 microprocessors are used in some personal computers, weapons systems and in some gaming machines.

(4) *Zilog Z8001*

The Z8001 has 23-bit addressing like the 68000. Several of its 16 16-bit work registers can be combined to give 32-bit working. It has been applied in the aerospace and missile industries.

(5) *National Semiconductor 16032*

The 16032 is the most popular of a series of 16-bit microprocessors offered by National Semiconductor. It possesses 16 M word addressing range and 32-bit register working.

(6) *Ferranti F100L*

This is the only British designed and manufactured microprocessor. Detailed information on its features is restricted due to the classified nature of its military applications.

BIBLIOGRAPHY

1. *Microcomputers and their Interfacing*, R. C. Holland, Pergamon, 1984.
2. *Getting Started with 8080, 8085, Z80 and 6800 Microprocessor Systems*, James W. Coffron, Prentice/Hall, 1984.
3. *Theory and Practice of Microprocessors*, K. G. Nichols and E. J. Zaluska, Crane, Russak and Company, 1982.

EXERCISES

1. The Digital Equipment Corporation PDP LSI 23 computer possesses a 16-bit word length and is used by Kent Process Control Limited to collect factory instrumentation information, and to display this information in a variety of forms on several colour display terminals. Which type of computer (classification) is this machine?
2. What functions are performed by an ALU?
3. Which CPU module examines the contents of the instruction register?
4. Which CPU module is a group of marker flags?
5. Which CPU module points to the memory address of the instruction which is to be obeyed next?
6. List the stages involved in the fetch/execute cycle for the Z80 instructions:

    (a) ADD A,7   ;Add 7 to the A register (accumulator)
    (b) RLC B   ;Rotate B register left circular
    (c) LD   HL,3000H   ;Load HL register-pair with
                       ;hex. 3000

7. What is the function of the Z80 control bus line RESET?

8. Describe the function of a stack pointer.

9. State the name of the addressing mode used in the destination in each of the following Z80 instructions, and describe the action of the instruction:

   (a) LD (HL),E
   (b) LD (904EH),B
   (c) LD L,H
   (d) LD (IY+3),2

10. Which of the following Z80 instructions does not produce the same result as the others if the accumulator initially contains 4?

    (a) RRA
    (b) SUB 1
    (c) DEC A

11. Do the following sections of programs for the Z80 have the same effect on memory location hex. 4000?

    (a) LD HL,4000H
        INC (HL)
    (b) LD A,(4000H)
        INC A
        LD (4000H),A
    (c) LD HL,4000H
        LD A,(HL)
        INC A
        LD (HL),A

12. Which instruction would you use to multiply a number held in the accumulator by 2?

13. Write a Z80 program in assembly language that adds the contents of memory locations hex. 6000 and 6500, and stores the answer in the E register.

14. Write a Z80 program in assembly language that stores 0 into 20 memory locations starting at hex. 0800.

15. State three advantages of the Z80 over the 8085 microprocessor.

16. State three advantages of the 6809 over the 6502 microprocessor.

17. Give three reasons why the Intel 8088 has been so widely applied in office computers.

CHAPTER 11

Memory Chips

11.1 INTRODUCTON

Memory ICs are MOS devices and are applied in computers to produce memory systems in which programs and data lists are held. Such systems are often termed "semiconductor memory" to distinguish them from the traditional core store systems, which employed small metal toroidal "cores" to store bits in magnetic form.

Microcomputer memory systems are normally sized from a few K bytes up to 64 K bytes, where:

$$1 \text{ K} = 2^{10} = 1024$$

There are two basic types of memory ICs:

(a) ROM (Read Only Memory)—once bits are stored initially they can only be read from the device,

(b) RAM (Random Access Memory)—bits can be read from and written to a RAM device.

Although RAM chips appear to be more useful than ROM chips, they suffer from the disadvantage that their stored bit pattern is lost when dc power is removed when the computer is switched off. Thus RAM memory is sometimes described as "volatile memory". Occasionally RAM memory systems are supported by back-up batteries.

Microcomputer circuits, from small applications like a washing machine controller up to large circuit applications like an office computer, possess some ROM and some RAM.

Within each category of memory device there are further sub-divisions. There are four different types of ROM, as follows:

(1) ROM—this is mask-programmed to store a specified bit pattern by the chip manufacturer;

(2) PROM (Programmable ROM)—this is manufactured with no stored bit pattern and can be programmed by the user;

(3) EPROM (Erasable PROM)—this can be erased by exposure to a UV (ultra violet) light source for typically 20 minutes and then re-programmed;

(4) EAROM (Electrically Alterable ROM)—this can be altered when it is connected in its final circuit; however erase and re-write time is long and EAROM circuits are cumbersome and expensive.

Frequently ROM, PROM and EPROM chips are pin-compatible, so that a program can be developed in an EPROM and then transferred to ROM or PROM to save chip cost when large-scale production runs are required.

The two subdivisions of RAM are:

(1) static RAM—stored bit pattern is retained as long as the dc supply to the chip is maintained,

(2) dynamic RAM—stored bit pattern is lost after a short period (typically 2 ms) unless the chip is "refreshed".

The most common examples of these memory chips are described in the following sections.

11.2 ROM AND RAM SYSTEMS

A typical ROM chip is shown in Fig. 11.1. The pin functions are:

11 address lines—11 two-state lines identify $2^{11} = 2048$ locations;

  8 data lines—each location stores 8 bits (1 byte);

  1 chip select—this activates the chip, i.e. the byte identified by the address lines setting is placed on the data lines.

There are additional dc supply pins (+5 V and 0 V).

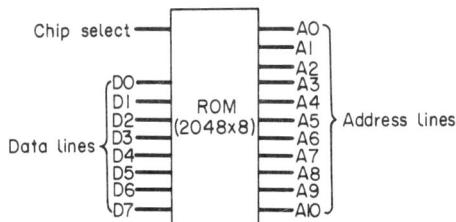

FIG. 11.1. Typical ROM chip.

The memory organisation of this chip is defined as:

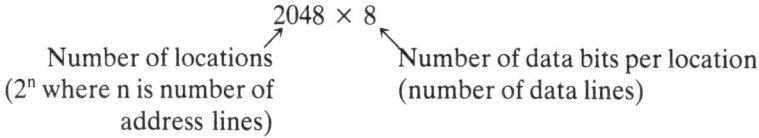

$$2048 \times 8$$

Number of locations (2^n where n is number of address lines)   Number of data bits per location (number of data lines)

PROM and EPROM versions of this chip are frequently pin-compatible. Therefore memory circuits for ROM, PROM or EPROM are identical.

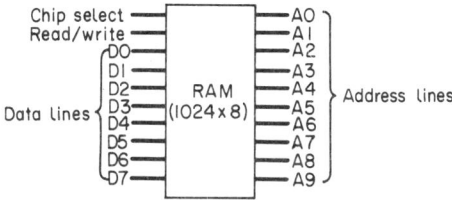

FIG. 11.2. Typical RAM chip.

Figure 11.2 illustrates a typical (static) RAM device. The pin functions are:

10 address lines—10 two-state lines identify $2^{10} = 1024$ locations
8 data lines—each location stores 8 bits (1 byte)
1 chip select—this activates the chip
1 read/write—this determines if bits are to be read from or written to the device.

Memory chips are connected to the CPU's buses in the manner shown in Fig. 11.3. The full 8-bit data bus is connected to each memory chip. Only the required number of address bus lines are connected to each memory chip.

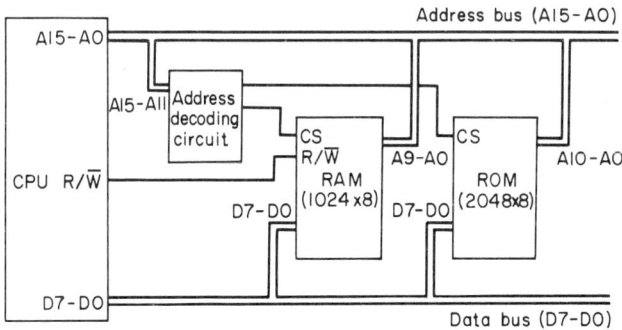

FIG. 11.3. Memory chips connected to CPU buses.

The only control bus signal required for this arrangement is the Read/write signal, which is connected to the RAM device. The address decoding circuit utilises the unused high-order address bus lines to generate chip select signals. Only one chip select signal is set at any time, so that only one memory chip can use the data bus. The effect of setting a memory IC chip select signal is to take the data signals out of the tristate "floating" state, i.e. they are connected electrically to the data bus.

11.3 ADDRESS DECODING

Section 5.4 described the operation of two common binary decoder circuits—a 2 to 4 decoder, and a 3 to 8 decoder. The former is available in the SN74139 TTL IC (two decoder circuits within the same package) and the latter is available in the SN74138 chip. Each of these circuits sets only one of its outputs (to logic 0) at any time; all other outputs are at logic 1. The output that is selected is determined by the binary code on the input connections.

Consider the application of a 2 to 4 decoder as an address decoding circuit which produces chip select signals for several memory chips, as shown in Fig. 11.4.

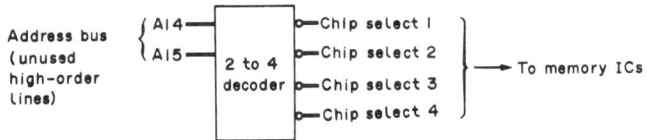

FIG. 11.4. Address decoding circuit ($\frac{1}{2}$ × 74139).

The truth table for this circuit is:

TABLE 11.1 TRUTH TABLE FOR
2 TO 4 DECODER

A15	A14	Chip select			
		4	3	2	1
0	0	1	1	1	0
0	1	1	1	0	1
1	0	1	0	1	1
1	1	0	1	1	1

Whenever the CPU implements a memory transfer operation, it sets the required memory address on the address bus. This address identifies one of the memory chips and a specific location within that chip. The start address of each chip is computed as follows:

	A15	A14	A13	A12	A11	A10	A9	A8	A7	A6	A5	A4	A3	A2	A1	A0
Chip select 1	0	0	X	X	X	X	X	X	X	X	X	X	X	X	X	X
Chip select 2	0	1	X	X	X	X	X	X	X	X	X	X	X	X	X	X
Chip select 3	1	0	X	X	X	X	X	X	X	X	X	X	X	X	X	X
Chip select 4	1	1	X	X	X	X	X	X	X	X	X	X	X	X	X	X

$$X = \text{don't care (assume 0)}$$

In this case the memory chips selected by the four outputs of this circuit possess hexadecimal start addresses as follows:

Chip select 1: Hex. 0000
Chip select 2: Hex. 4000
Chip select 3: Hex. 8000
Chip select 4: Hex. C000

If a program instruction demands that the contents of memory location hexadecimal 403A are read into the CPU, then Chip select 2 is set, and location 3A within the selected chip is accessed.

The full memory circuit diagram of a microcomputer is shown in Fig. 11.5.

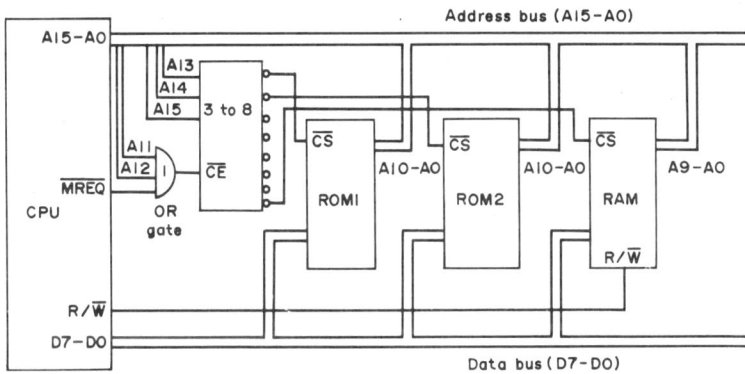

FIG. 11.5. Typical microcomputer memory circuit.

In this case a 3 to 8 decoder is used, although a 2 to 4 decoder would have been sufficient. However five spare chip select signals are available to allow extra memory chips to be added. The decoder itself possesses a chip select/enable function ($\overline{CE}$) and this signal is generated in an OR gate from a combination of the unused address bus lines A11 and A12, and the control bus line $\overline{MREQ}$ (set to 0 by the CPU during memory transfers). All three signals must be set to 0 to generate a 0 to activate the decoder chip. Therefore the start and end addresses of the three memory chips are computed as follows:

	A15	A14	A13	A12	A11	A10	A9	A8	A7	A6	A5	A4	A3	A2	A1	A0	Hex.
ROM1(start):	0	0	0	0	0	0	0	0	0	0	0	0	0	0	0	0	0000
ROM1(end) :	0	0	0	0	0	1	1	1	1	1	1	1	1	1	1	1	07FF
ROM2(start):	0	0	1	0	0	0	0	0	0	0	0	0	0	0	0	0	2000
ROM2(end) :	0	0	1	0	0	1	1	1	1	1	1	1	1	1	1	1	27FF
RAM(start) :	1	1	1	0	0	X	0	0	0	0	0	0	0	0	0	0	E000
RAM(end) :	1	1	1	0	0	X	1	1	1	1	1	1	1	1	1	1	E3FF

3 to 8 decoder	OR gate (to activate 3 to 8 decoder)	Connected to memory chips

The truth table for the 3 to 8 decoder can be verified by reference to section 5.4.

This memory system produces the "memory map" for the microcomputer as shown in Fig. 11.6. a memory map is a convenient way of showing how the full 64 K of microcomputer memory is utilised. Notice the large unused areas of memory space in this application.

The reader may like to exercise his understanding of the techniques for calculating memory addresses, and therefore constructing a memory map,

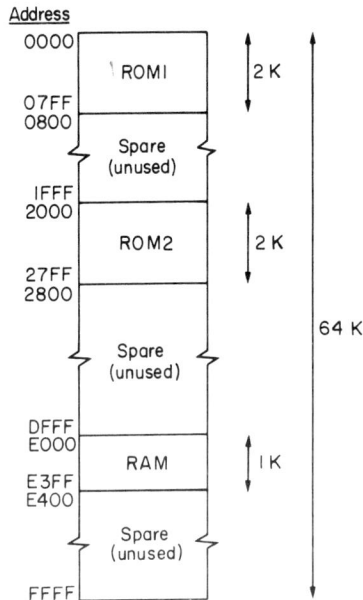

FIG. 11.6. Memory map for Fig. 11.5.

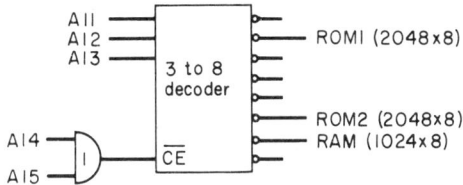

Fig. 11.7. Sample address decoding circuit.

from a circuit diagram by repeating the above analysis for the address decoding circuit shown in Fig. 11.7. The answers should be:

	Start	End
ROM1	0800	0FFF
ROM2	2800	2FFF
RAM	3000	33FF

Some circuit designers use gating networks in place of 2 to 4 and 3 to 8 decoders in order to generate chip select signals from unused address bus lines. Normally it is simpler to use decoder ICs.

11.4 TYPICAL ROM—2316

The 2316 ROM IC, which is illustrated in Fig. 11.8, is one of the most common ROM devices, and it offers a memory organisation of:

2048 × 8   (2048 locations each containing 8 bits)

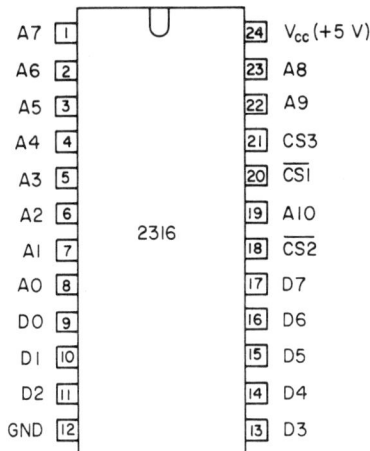

Fig. 11.8. 2316 ROM.

There are 11 address lines, where:

$$2^{11} = 2048$$

Notice that there are three chip select signals, such that the device is only activated (data pins D0 to D7 are taken out of the floating state) when:

CS1 = logic 0
CS2 = logic 0
CS3 = logic 1

The access time is typically 500 ns or less, and the power consumption is 0.5 W (watt).

Notice that the last two digits of the serial number of this 2316 chip are 16, and they are used to indicate the number of storage bits in units of 1 K— 16 K bits in a memory organisation of 2048 × 8. The ROM chip which offers double this storage capacity is the 2332—32 K bits in a memory organisation of 4096 × 8.

11.5 TYPICAL EPROM—2716

The widespread popularity of the 2716 EPROM is because it offers a memory capacity of 2 K bytes (2048 × 8), which is ideal for many small microcomputer applications. Its pin functions are shown in Fig. 11.9. Notice that it is designed to be pin-compatible with the 2316 ROM, so that an application prototype can be based on a 2716, which gives the flexibility of being able to be re-programmed, but can then be replaced by a 2316 for large

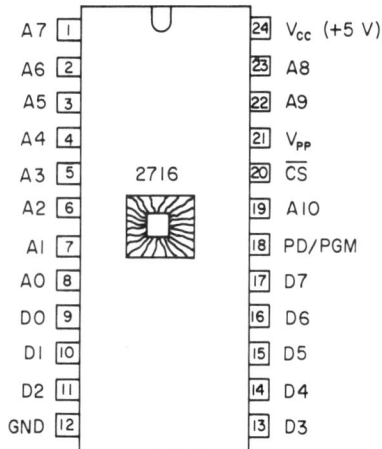

FIG. 11.9. 2716 EPROM.

production runs to save cost. The 2716 is typically four times the cost of a 2316.

The three chip select pins on the 2316 are replaced by a single chip select signal (pin 20) plus the following two signals:

(a) V_{PP}, which is set to $+24$ V when the device is programmed initially; this signal must be set to logic 1 ($+5$ V) when the chip is placed in its final application circuit;

(b) PD/PGM (power down/programming), which is set to logic 1 for a short period when each byte is programmed into the device (this is the "programming" function); this signal must be set to logic 1 when the chip is addressed in its final application circuit so that it operates in the "power down" mode, and it must be pulsed to logic 0 when a byte is read from the device.

Figure 11.10 shows the timing diagram for both the initial programming of a 2716 EPROM and the reading of a byte from the device when it is installed in its final circuit. Programming one byte takes 50 ms, and so the entire process of programming a complete 2 K byte 2716 is:

$$2048 \times 50 \text{ ms} = 100\text{s approximately}$$

All signals switch between logic 0 (0 V) and logic 1 ($+5$ V) except V_{PP} in (a). The crossover effect shown on the address lines denotes that each of the lines can change from 0 to 1 and vice versa. The data lines are shown in the floating state except when the data signals are active, e.g. when a data byte is read out of the device in (b).

A blank EPROM contains all 1s when supplied by the manufacturer. Similarly after erasure by a UV (ultra violet) light source each memory byte is forced to contain hexadecimal FF. An EPROM eraser is simply a container into which EPROMs (normally supported on a tray) can be inserted. Erasure time is typically 20 minutes. Care must be taken to prevent eye-contact with the UV lamp—the eye can be damaged. The erasure window on the top of an EPROM must be covered with an opaque label after the device is programmed to prevent erasure caused by fluorescent lighting or sunlight (300 hours of continuous exposure to sunlight can erase an EPROM).

Other common EPROMs are:

(1) 2708—1 K bytes (1024 × 8)
(2) 2732—4 K bytes (4096 × 8)
(3) 2764—8 K bytes (8192 × 8)

Access times are typically 400 ns and power consumption is typically 0.8 W.

(a) Initial programming (2 bytes only shown)

(b) Reading a byte

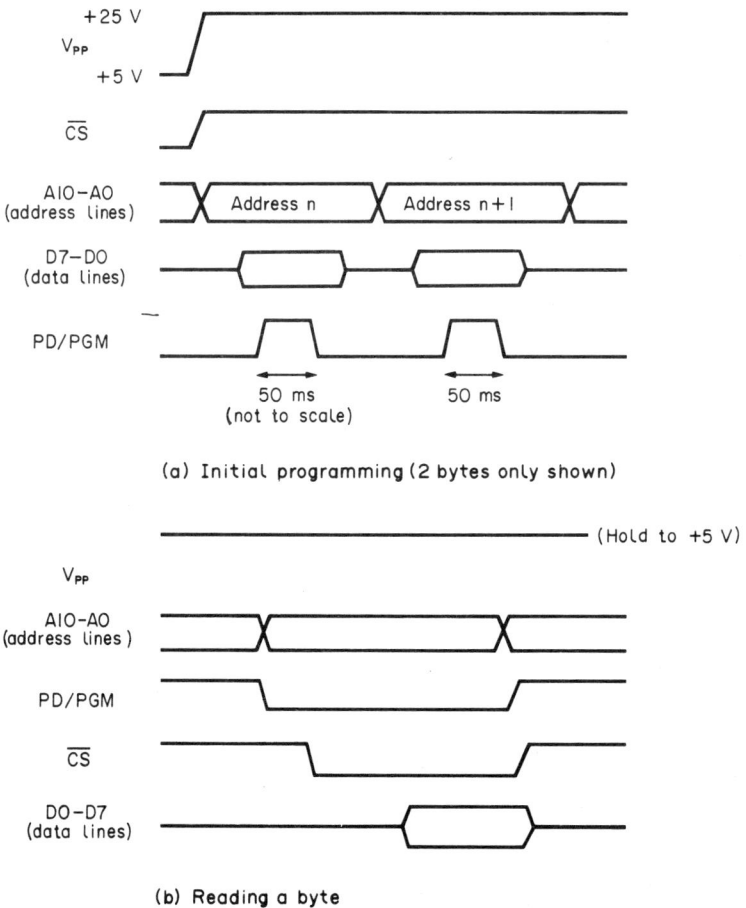

Fig. 11.10. Timing diagram for 2716 EPROM.

11.6 TYPICAL STATIC RAM—2114

Although 8-bit static RAM devices are available, the most common static RAM IC is probably the 2114, which is a 4-bit device. Its memory organisation is 1024×4, i.e. it possesses 1024 locations, each containing 4 bits (1 "nybble"). Its pin functions are illustrated in Fig. 11.11. It possesses 10 address lines, where

$$2^{10} = 1024$$

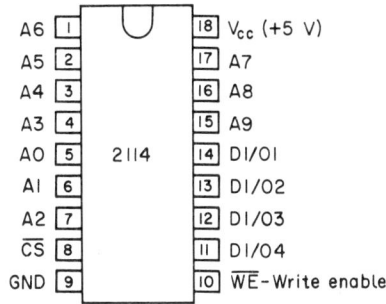

FIG. 11.11. 2114 Static RAM.

There are only four data lines (DI/O1 to DI/O4), which are bi-directional. The $\overline{WE}$ (NOT Write Enable) selects the direction of data transfer, and this signal pin is connected to the CPU's control bus line (R/$\overline{W}$ for the 6502 and 6809) which identifies memory read or write.

Access time for the 2114 is 300 ns and power consumption is typically 1 W.

Byte storage is achieved by connecting two 2114s to the CPU's data bus, with one chip storing the least significant nybble and the other chip storing the most significant nybble, as shown in Fig. 11.12.

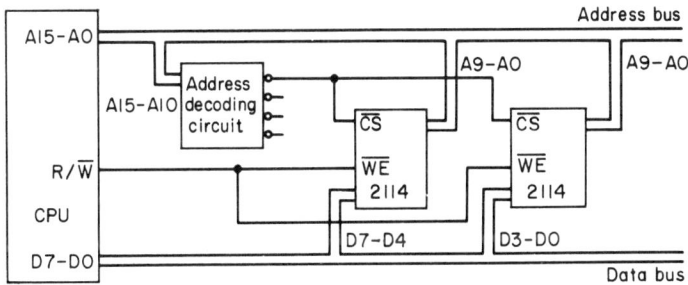

FIG. 11.12. Byte storage using 4-bit RAM devices.

11.7 TYPICAL DYNAMIC RAM—4116

Dynamic RAM devices provide high capacity storage but with reduced chip area compared with static RAMs. They provide more cost-effective systems when large memory arrays, e.g. 16 K or more, are required.

The most popular dynamic RAM IC is the 4116, which is illustrated in Fig. 11.13. It possesses a memory organisation of 16384 × 1, i.e. it has 16384

```
        -5 V- V_BB  [1]   ⎰‾‾⎱  [16]  V_SS - GND
       Data in - D_in  [2]        [15]  C̄ĀS̄ - Column select
    Write enable - WRITE  [3]     [14]  D_out - Data out
       Row select - RAS  [4]      [13]  A6
                    A0  [5]  4116  [12]  A3
                    A2  [6]        [11]  A4
                    A1  [7]        [10]  A5
            +12V - V_DD  [8]        [9]  V_CC - +5 V
```

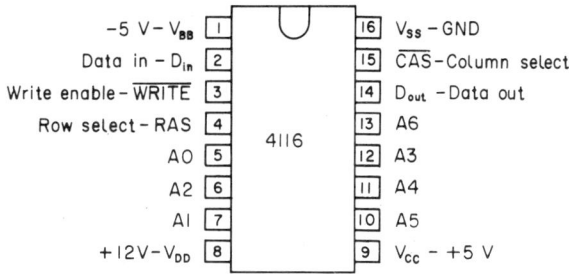

FIG. 11.13. 4116 Dynamic RAM.

locations each containing 1 bit. Eight such devices must be connected to a CPU's data bus to give 16 K byte storage. Notice the following features:

(a) Three power supplies are required: $+5$ V, -5 V and $+12$ V.

(b) Two separate data in and out signal lines exist—these must be combined externally before connection is made to one of the bi-directional data bus's lines.

(c) The chip select function is achieved using the $\overline{\text{RAS}}$ signal.

(d) Only one-half of the expected address line connections are available—a 16 K memory device should possess 14 address lines, whereas only 7 exist on the chip. This incompatibility is overcome by arranging that an external circuit presents the lower order 7 address lines (A6–A0) on the address pins firstly, together with $\overline{\text{RAS}}$, and then approximately 100 ns later the higher order address lines (A13–A7) are placed on the same address pins, together with $\overline{\text{CAS}}$ ($\overline{\text{RAS}}$ must remain set).

Dynamic RAM loses its bit pattern after 2 ms unless the stored data are refreshed. This refresh is achieved by simply addressing the bottom 128 addresses, i.e. setting every possible bit pattern on A0 to A6, faster than once every 2 ms (together with $\overline{\text{RAS}}$). Normally an external 7-bit counter is used to step through each of these address settings when the CPU is not using the dynamic RAM chip for memory transfers.

Access time for the 4116 is 200 ns and power consumption is typically 0.5 W.

A simplified diagrammatic representation of a memory system that supports a Z80 CPU with 16 K bytes of dynamic RAM using 4116 ICs, plus the facility to add 48 K bytes additional memory, is shown in Fig. 11.14. A 7-bit selector circuit (using 2 off 74157) passes out either the least significant half of the 14 address bus lines (A6–A0) which feed to all dynamic RAMs, or a 7-bit refresh count (using 2 off 7493). The counter is incremented by the

A6–A0, or
refresh address

A0
A6

A7
A13

Selector
(2 off
74157)

Selector
(2 off
74157)

A6–A0 firstly (together with
RAS set), or refresh address,
A13–A7 secondly (together with
CAS and RAS set).

MREQ

RFSH — 7-bit
counter (2
off 7493)

MREQ

Delay
100 ns
74221

CAS

RAS

A14
A15

2 to 4
decoder
74139

To other
banks of
8 4116s

A6–A0

A6–A0

RAS
CAS

4116

WRITE

D_in   D_out

RAS
CAS

4116

WRITE

D_in   D_out

D_in 7   D_out 7   D_in 0   D_out 0

Bidirectional
data bus

Buffer
74244

Separate
unidirectional
data bus in and
data bus out

D0
D7

Buffer
74244

RD

WR

8 4116 dynamic RAM ICs
(only 2 shown for simplicity)

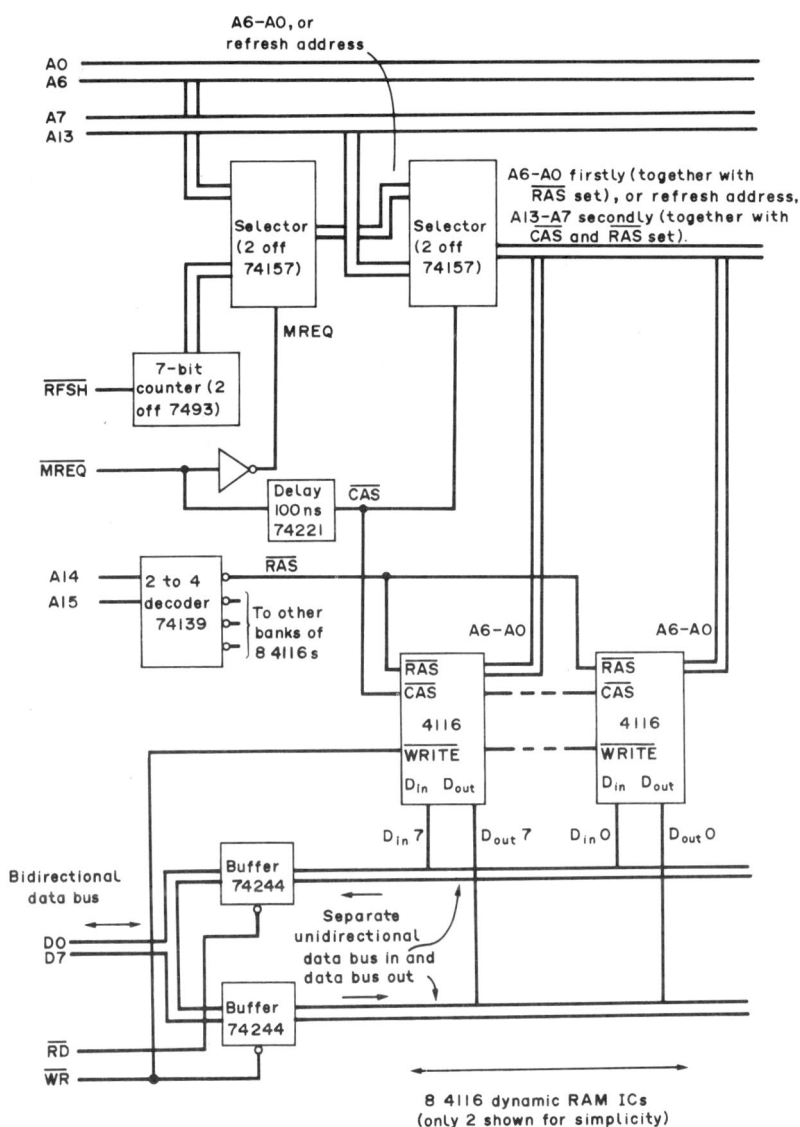

FIG. 11.14. Simplified schematic of 16 K byte dynamic RAM using 4116 ICs.

CPU's $\overline{\text{RFSH}}$ pulse which is generated when the CPU is inactive between instructions. A second selector passes either the least significant half of the address lines (A6–A0)—or the refresh address—or the most significant half of the address lines (A13–A7) to the 7 address pins on *all* dynamic RAMs. The top two unused address lines A15 and A14 are used for address decoding. Each output of the decoder selects one of four banks of 16 K of memory—constructed using 8 off 4116 ICs. Only one bank is shown, and only two of the eight chips are illustrated for simplicity. The CPU's data bus is split into two unidirectional data buses using two tri-state output buffers (74244).

This arrangement of dynamic RAMs is common with home computers and business computers.

BIBLIOGRAPHY

1. *Microelectronic Systems Level 111*, D. J. Woollons, TEC/Hutchinson, 1982.
2. *Introduction to MOS LSI Design*, J. Mavor, M. A. Jack and P. B. Denyer, Addison-Wesley, 1983.
3. *Study notes for Technicians: Microelectronic Systems Level 3*, R. C. Holland, McGraw-Hill, 1983.

EXERCISES

1. What does the description "volatile" memory mean?
2. Which type of read-only memory can be programmed more than once by the user?
3. What is the name of the signal that activates a memory chip (takes its data signals out of the "floating " state)?
4. List the pin functions of an 8 K byte ROM chip.
5. List the pin functions of a 512 × 4 RAM chip.
6. If an address decoder chip is not available, how would you design a microcomputer memory circuit, which has two memory chips, such that only one memory device is selected at any time?
7. Design a memory circuit using a 2 to 4 decoder which generates chip select signals to select four 1024 × 8 ROM chips with start addresses hex. 0000, 0400, 0800 and 0C00.
8. What are the start and end addresses of the memory chips fed by the following address decoding circuit?

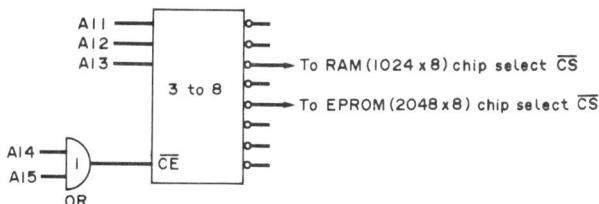

A11, A12, A13 — 3 to 8 decoder
To RAM (1024 x 8) chip select $\overline{\text{CS}}$
To EPROM (2048 x 8) chip select $\overline{\text{CS}}$
A14, A15 — OR — $\overline{\text{CE}}$

9. Describe the functions of the two pins which a 2716 EPROM possesses and which are the only signals that are not compatible with a 2316 ROM.
10. What is the memory organisation of a 2708 EPROM?
11. If the least significant four bits are mis-read from every location in the memory system shown in Fig. 11.12, which chip is the most likely cause of the fault?
12. What is the principal reason for using dynamic RAM memory arrays in place of static RAM? Suggest other advantages.
13. How is dynamic RAM "refreshed"?
14. Examine Fig. 11.14. What is the start address of a 16 K byte bank of RAM fed by a $\overline{RAS}$ signal from the bottom output of the 74139 chip?

CHAPTER 12

Input/Output Chips

12.1 GENERAL INPUT/OUTPUT CIRCUITS

Input/output chips connect to a microprocessor's three buses (address, data and control) in exactly the same manner as memory chips. Figure 12.1 illustrates the way in which three input/output chips are connected into a microcomputer circuit. An address decoding circuit (2 to 4 decoder in this case) generates only a single chip select signal at any time, so that only one input/output chip can be activated by a program instuction. The control bus signal IO/$\overline{\text{M}}$ is shown connected to the chip select pin of the decoder to ensure that the decoder is active only when input/output transfers occur. The same control signal is inverted and passes to the chip select pin of a further decoder chip which activates one of several memory chips. This arrangement is necessary to ensure that only one decoder chip is selected; thus a memory chip and an input/output chip cannot be selected together. Therefore

F1G. 12.1. Simplified microcomputer input/output circuit.

146

input/output chips and memory chips can have the same addresses. The CPU sets the IO/$\overline{\text{M}}$ signal appropriately when it is performing either an input/output transfer or a memory transfer. The alternative arrangement is to share memory and input/output chip select signals from the same address decoder. This is termed "memory mapped input/output", and in this system memory transfer instructions are used to perform input/output operations.

Data are transferred to and from input/output chips in 8-bit (byte) form—the data bus connection to the CPU is 8-bits and normally the input/output signals to remote devices and peripherals are grouped in 8 bits. Every microprocessor manufacturer supports his CPU chip with a range of MOS input/output chips. Commonly these are nearly as complex as the CPU itself. The two most common chips are the parallel input/output (PIO) and the serial input/output (UART) devices, which are described in following sections. Additionally most manufacturers offer specialised input/output chips to service common peripheral equipment, e.g. floppy disk and video generator to a CRT. Again these devices are described in the following sections.

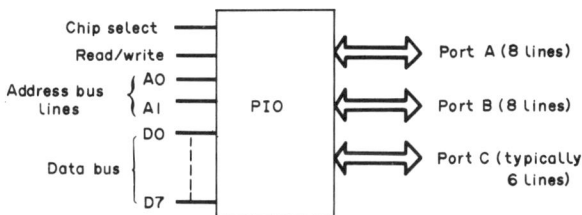

Fig. 12.2. PIO (typical).

12.2 PARALLEL INPUT/OUTPUT (PIO)

A large number of input/output connections to microcomputers pass through parallel input/output (PIO) chips. The majority of PIOs are 40-pin devices and offer two or three "ports" for external connections. A port is a set of 8 external signals, which can be either input or output. Figure 12.2 shows the pin functions of a typical PIO. This device possesses three ports, although the third port (Port C) has only 6 signal lines—this is common with three-port PIOs. The chip select signal activates the device (takes the data bus connections out of the floating state) in the same manner as for memory chips. Similarly the read/write signal selects the direction of data transfer (for input or output instructions). There are two address bus lines connected, to give a total of 4 addresses as follows:

A1 A0

0	0	—Port A
0	0	—Port B
1	0	—Port C
1	1	—Control register

If chip select is generated by an address decoder when A2 = 1 and A3 = 0, these four addresses are: hexadecimal 04, 05, 06 and 07

The control register is an addressable circuit within the chip and it makes the device "programmable", i.e. each port can be programmed to be either input or output in direction. Therefore the software must output a control byte to this address to select the directions of the ports before data bytes are transferred in or out via the ports.

Typical applications of output ports are parallel connection to:

(a) 7-segment numerical display, with eighth bit for decimal point (see section 5.5),
(b) parallel drive printer,
(c) D/A chip to generate an analogue signal,
(d) 8 single-bit circuits, e.g. indicator LEDs, transistor drive to large electrical load (motor, heater), etc.

Typical applications of input ports are parallel connection to:

(a) keyboard (8-bit code representing key pressed),
(b) A/D chip to enable an analogue signal to be read into the micro-computer,
(c) 8 single-bit signals, e.g. push-button closures, electrical contact closures.

Details of the Zilog Z80 PIO are given in section 12.6. Frequently the functions of another common input/output chip (the CTC) are combined within a PIO; the CTC is described in section 12.4.

An alternative chip to a programmable two- or three-port MOS PIO is a non-programmable 8-bit register, which is available in the TTL 7400 series. There are several devices that can operate as input ports, e.g. SN74240, SN74241 and SN74244, which are detailed in Appendix A and possess the following pin functions:

8 input lines (connect to external signals),
8 output lines (connect to data bus)—tri-state,
2 enable lines (connect to an output of an address decoder, e.g.
2 to 4 decoder, to take the output lines out of the "floating" state).

A flexible device that can operate either as an input port or as an output port is the SN74373, which is shown in Fig. 12.3. The device is selected and

Key

EL = Enable latch
OE = Output enable (inverse logic)

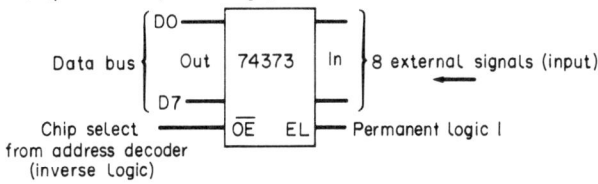

Data bus { DO — Out | 74373 | In — D7 — } 8 external signals (input)

Chip select ———— $\overline{OE}$ EL — Permanent logic I
from address decoder
(inverse logic)

(a) Connected as input port

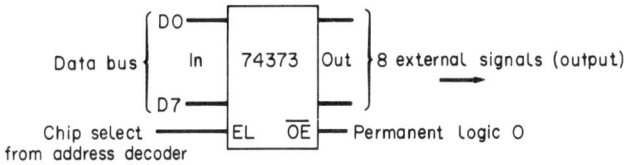

Data bus { DO — In | 74373 | Out — D7 — } 8 external signals (output)

Chip select ———— EL $\overline{OE}$ — Permanent logic O
from address decoder

(b) Connected as output

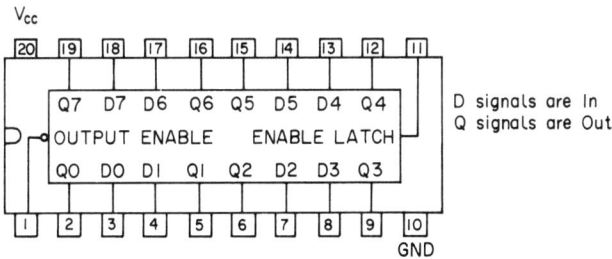

V_{cc}

| 20 | 19 | 18 | 17 | 16 | 15 | 14 | 13 | 12 | 11 |

| Q7 | D7 | D6 | Q6 | Q5 | D5 | D4 | Q4 |

OUTPUT ENABLE   ENABLE LATCH

| Q0 | D0 | DI | QI | Q2 | D2 | D3 | Q3 |

| 1 | 2 | 3 | 4 | 5 | 6 | 7 | 8 | 9 | 10 |
                                        GND

D signals are In
Q signals are Out

(c) Pin functions

Fig. 12.3. Non-programmable port—the SN74373 (octal transparent latch).

passes 8 bits from its In lines to its Out lines (and staticises these bits) only when Enable Latch is set (to 1) and $\overline{\text{Output Enable}}$ is set (to 0). In (a) the chip is wired into a circuit as an input port, and 8 external signals pass onto the data bus when $\overline{OE}$ (fed from an address decoder circuit) is set to 0. In (b) the chip is wired into a circuit as an output port, and the setting on the data bus passes out to 8 external devices when EL (fed from an address decoder circuit) is set to 1. Notice how an output from an address decoder circuit, e.g. 2 to 4 decoder, is used to activate the chip in each case.

An even simpler arrangement for output signals is achieved using a simple

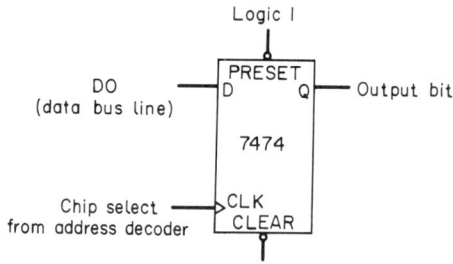

(a) Single-bit output (D-type flip-flop)

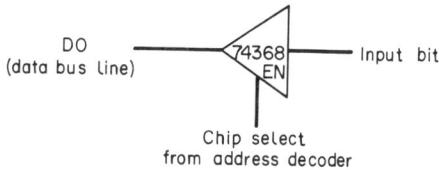

(b) Single-bit input (3-state buffer)

FIG. 12.4. Single-bit input/output circuits.

flip-flop if only a single-bit output signal is required, as shown in Fig. 12.4. A single-bit input circuit using a 3-state buffer is also shown in Fig. 12.4. Once again each of these circuits is activated only when an output from an address decoder circuit is generated.

12.3 SERIAL INPUT/OUTPUT (UART)

Serial input/output transfers are achieved by passing data items in the form of bytes along a single conductor for each direction of transfer. The eight bits are pulsed along the conductor one after the other, and so serial data transfer is slower than parallel transfer. However it does reduce cable conductors, and this can yield cost savings. Serial transmission is applied to connect microcomputers to VDUs, other computers and printers (although parallel connection is frequently applied with printers).

The IC that performs parallel to serial conversion for output, and serial to parallel conversion for input, is the UART (Universal Asynchronous Receiver Transmitter). In the same way that a microprocessor manufacturer

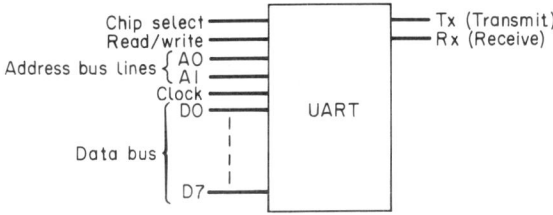

FIG. 12.5. UART (typical).

supports his CPU chip with a PIO, he normally offers a UART which can be simply connected to the CPU buses. The pin functions of a typical UART are shown in Fig. 12.5. The chip select and read/write signals have the same functions as with a PIO or RAM chip. The clock signal is required to trigger the parallel to serial (and vice versa) conversion processes. The two address lines typically give the following addresses on the chip:

A1   A0
0   0 —Tx (serial output channel)
0   1 —Rx (serial input channel)
1   0 —Control register
1   1 —Status register

The status register is used to indicate if a character byte has been received, or if a transmitted byte has been cleared. The control register is used to "initialise" the device by software in much the same way as a PIO is initialised. Initialising a PIO involves selecting the directions (output or input) of the ports, whilst initialising a UART involves setting the transmission speed and the variable modes of operation which are listed in the RS232-C specification. RS232-C is a world-standard for serial data transmission, and its main features are:

(a) *Signal Waveform*

The RS232-C signal waveform is shown in Fig. 12.6. Seven data bits plus 1 parity bit are used normally. The full 8-bit character is framed by a start bit

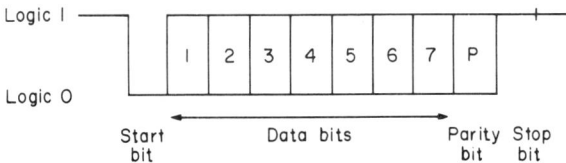

FIG. 12.6. RS232-C signal waveform for 8-bit character.

ICM-K

(logic 0) and a stop bit (logic 1)—sometimes 1½ or 2 stop bits are used. Normally the ASCII (American Standard Code for Information Interchange) character set, which is listed in Appendix B, is applied. Occasionally 5 or 6 data bits are selected in place of 7. The parity bit is used to provide a check on successful transmission of the data bits, e.g. if even parity is selected the parity bit is set to 1 or 0 to make the overall number of 1s an even number. Odd parity is the converse.

(b) *Signal Levels*
   The specified accceptable signal level ranges are:

   Logic 1 = −12 V approximately (−3 V to −25 V)
   Logic 0 = +12 V approximately (+3 V to +25 V)

A variation of the RS232-C standard is the RS422 standard, in which the logic levels are 20 mA and 0 mA (for 1 and 0 respectively) current signals. The generation of these signals requires additional circuitry beyond the UART, but this is often justified because RS422 current signals survive long-distance data links better than RS232-C voltage signals.

(c) *Transmission Speeds*
   The following transmission speeds are applied (1 baud = 1 bit/second):

   110, 300, 600, 1200, 2400, 4800, 9600, 19200 baud

For example, data link from a microcomputer to a VDU set to run at 4800 baud transmits 480 characters per second (assuming 7 data bits, 1 parity bit and 1 stop bit).

(d) *Interconnection Plug*
   A 25-pin "D"-type plug is required, and the pin functions are:

   Tx (Transmit)—pin 2
   Rx (Receive)—pin 3
   0V (signal ground)—pin 7

Other handshaking signals that can be applied are:

   RTS (Request to Send)—pin 4
   CTS (Clear to Send)—pin 5

The UART that generates and receives the RS232-C signal must firstly be initialised by software (by software sending a control byte to the control register) to select the following options:

(1) baud rate (for both transmit and receive),
(2) number of data bits (5, 6 or 7),
(3) number of stop bits (1, $1\frac{1}{2}$ or 2),
(4) parity (odd or even).

Figure 12.7 shows a typical connection arrangement for a microcomputer data link to a VDU. If the data link passes through the telephone network a modem is placed at each end of the link (see section 9.5) to convert the RS232-C pulses into sinewaves, and vice versa.

Some UARTs are termed USARTs (Universal Synchronous and Asynchronous Receiver Transmitter) because they possess the further option of synchronous transmission. In this mode a start bit and stop bit are not used, and the timing transfer of each bit from the transmitter circuit to the receiver circuit is synchronised by means of a common clock signal. Synchronous transmission is rarely used.

More details of the application of a UART are given in section 12.6.

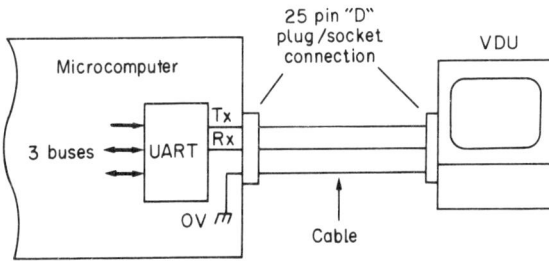

FIG. 12.7. RS232-C link to VDU.

12.4 COUNTER/TIMER CIRCUIT (CTC)

A counter/timer circuit is often offered by a CPU manufacturer as a separate IC, or sometimes it is included within a PIO chip. Basically it is simply a counter that can be loaded from the CPU, or read into the CPU, using the data bus, in the manner shown in Fig. 12.8. The circuit also generates a Count Complete signal, which is set when the counter reaches zero during count-down. Two common applications of this circuit are:

(a) *To generate pulses of a fixed repetition rate*
A program instruction sends out a binary count to the CTC. This count is loaded into the counter, which proceeds to count down using fixed frequency

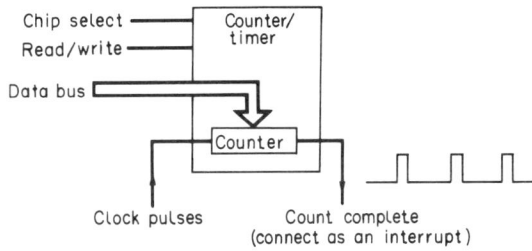

(a) Generate pulses of fixed repetition rate

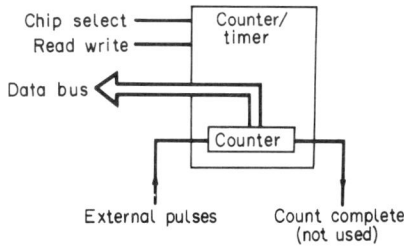

(b) Count external pulses

Fig. 12.8. Counter/timer circuit (CTC).

clock pulses. This clock signal is normally the CPU clock signal, or a derivative of it, and it possesses a stable repetition rate due to the crystal-controlled feature of the CPU clock generator circuit. When the count reaches zero, the Count Complete signal is set. If the CTC has been initialised to reload the counter with the original count, the cycle repeats and a repetitive pulse stream is generated on the Count Complete line. This signal is commonly connected to one of the CPU's interrupt lines, so that an interrupt program is called at regular intervals. In this way the interrupt program can update a time-of-day count in memory. Other programs can use this time-of-day for display purposes or to call other programs at specific times. The pulse stream on the Count Complete line can also be used as a clock signal to a UART.

An alternative application of a CTC in this count-down mode is when it is required to generate a fixed length delay in a program. In this case the CTC is not initialised to reload the counter, and it only generates one Count Complete signal. This signal can be polled (continuously read by software until it changes state), or it can be connected to an interrupt line.

(b) *To count external pulses*

In this case the CTC is normally initialised to count up rather than count down, and the counter value is read into the CPU by software whenever it is required to check the count attained. The counter can be reset at any time by an output instruction. A typical application is when external events, e.g. vehicles passing a checkpoint, need to be counted. This method avoids the software overhead of continuously scanning a single-bit input to detect each time it changes.

Notice that in both of the above modes of operation the CTC must be initialised, or "programmed", in much the same way as a PIO and UART. In the case of the CTC the device is initialised to either count up or count down, to reload itself and repeat the countdown, to reset itself (to zero) and even to divide the incoming Clock pulses by a binary value before changing the counter.

Some CTC chips possess more than one counter circuit, e.g. the Zilog CTC has four separate counter systems. It is described more fully in section 12.6.

12.5 SPECIAL FUNCTION INPUT/OUTPUT CHIPS

The PIO, UART and CTC described so far are flexible devices which are used for a wide range of input/output applications. Several other chips have been designed to drive just one type of peripheral as a result of the common use of that peripheral with microcomputers, e.g. keyboards, floppy disks and television monitors.

(a) *Keyboard Encoder*

The conventional method of reading the manual operation of a keyboard involves reading in blocks of key settings through an input port and then checking by software if any key is pressed. This must be repeated on a regular basis, e.g. 10 times every second, to avoid missing the manual operation of one of the keys. This considerable software overhead can be avoided by the use of a keyboard encoder chip, which itself performs the regular scanning function and notifies the CPU when a key is pressed.

Figure 12.9 shows the use of a typical keyboard encoder IC. The chip monitors a 4×4 keyboard—larger ICs are offered to handle 8×8 keyboards. The chip sets each of the column lines in turn and reads in on 4 row lines the settings of a block of 4 keys for each column. It then repeats the process continuously. If any key is pressed it sets the Data Available signal, which commonly is connected to one of the CPU's interrupt lines. The CPU

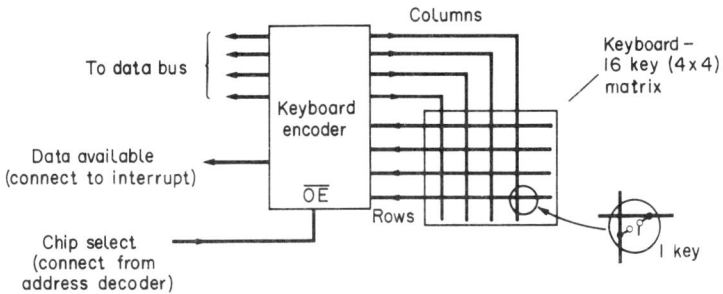

(a) Connections

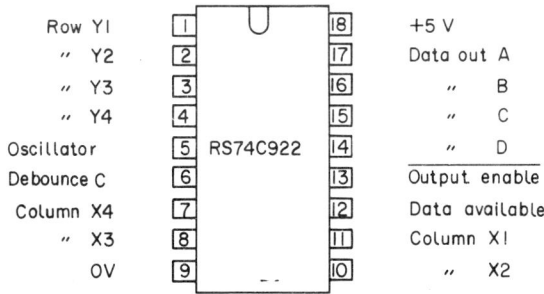

Row YI	1		18		+5 V
" Y2	2		17		Data out A
" Y3	3		16		" B
" Y4	4		15		" C
Oscillator	5	RS74C922	14		" D
Debounce C	6		13		Output enable
Column X4	7		12		Data available
" X3	8		11		Column XI
OV	9		10		" X2

(b) Pin functions of RS74C922

FIG. 12.9. Keyboard encoder IC (for 16-key keyboard).

then obeys an interrupt program which reads in along the data bus (only 4 lines used) a 4-bit code which identifies which key is pressed. The pin functions in diagram (b) show that a capacitor must be connected to each of pins 5 and 6 to enable the internal clock and debounce circuit.

(b) *Floppy Disk Controller*

The floppy disk has become the most popular device that is used for bulk storage of programs and data files with microcomputers. The portable nature of the interchangeable floppy disk makes it an extremely flexible and cost-effective device for changing the set of programs which a microcomputer runs. The replacement of one floppy disk for another within the floppy disk drive unit can change the function of an office microcomputer from a general commercial role (e.g. sales ledger, payroll) into the role of a scientific design machine (e.g. process and display test data).

The physical appearance of an 8″ floppy disk is shown in Fig. 12.10. The

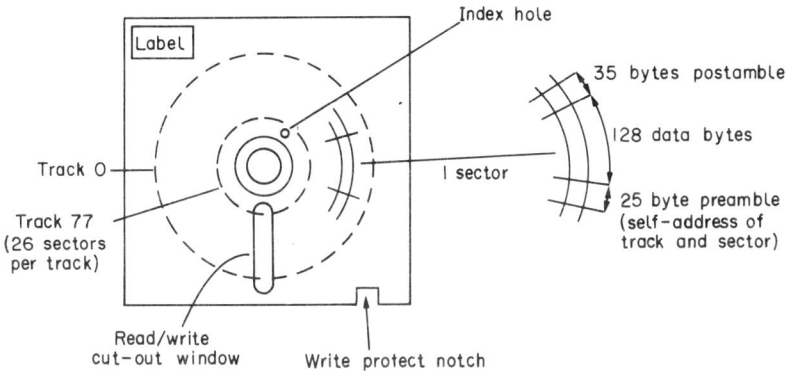

Fɪɢ. 12.10. Floppy disk (8″).

plastic disk is coated with magnetic material, and it rotates inside a paper envelope. The read/write head comes into contact (via a protective felt pad) with the disk surface through a read/write cut-out window in the envelope during read and write operations. The disk surface is divided up according to the world-standard IBM 3740 format into tracks (77 for an 8″ disk, fewer for the alternative smaller $5\frac{1}{4}$″ disk) and sectors (26 per track). Each sector is 128 bytes (for single-density recording) or 256 bytes (for double density recording) and each sector has a preamble (including its own address) and postamble (simply a data gap). Disks can be single-sided or doubled-sided (if two read/write heads are used).

A single IC can provide the complete interface between a microprocessor and a floppy disk drive unit, and the principal pin functions of such a chip are shown in Fig. 12.11. The device fits onto the CPU buses in the usual manner—the data bus, chip select and WE (write enable to select direction of data transfer) are connected as normal. An interrupt signal is generated

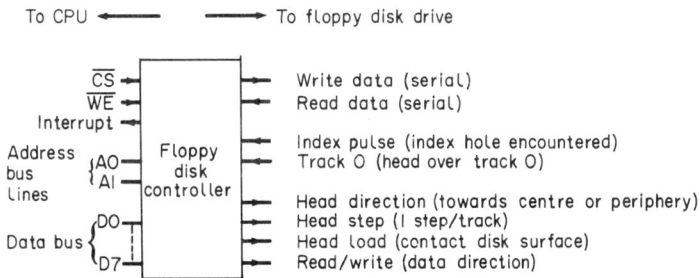

Fɪɢ. 12.11. Floppy disk controller.

and is normally set when a complete sector read or write operation is completed. The two address lines allow the following four registers to be addressed:

Data register,
Track register,
Sector register,
Status/Command register.

Data bytes are written to, and read from, the disk in serial form due to the rotating action of the disk. The Index pulse indicates when the index hole is encountered, and Track 0 indicates when the read/write head is over track 0. The three Head control signals on the device move the head by stepper motor drive to the required track (1 step/track), and then place the head in contact with the disk surface when the required sector has rotated beneath the head.

Typical floppy disk controller chips are the FD1771B-1 and the WD2797-02.

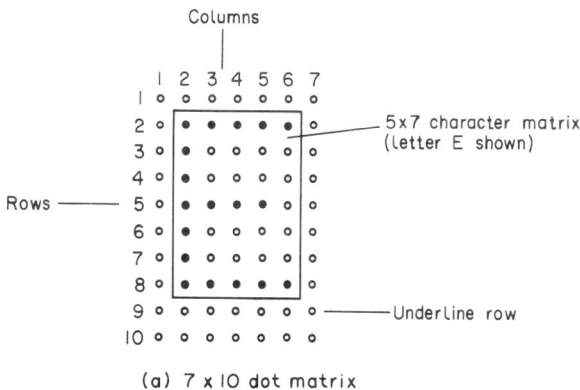

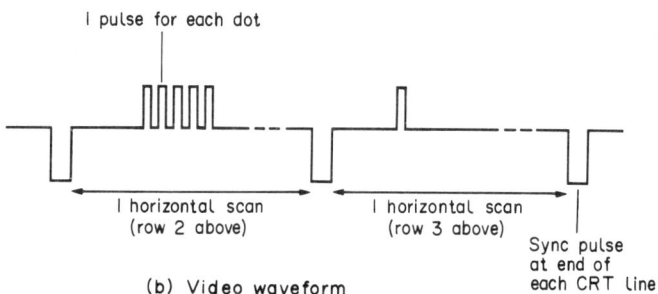

Fig. 12.12. Character display on a CRT.

(c) *Video Generator*

The interface between a microcomputer and an operator VDU (or "terminal") is normally by serial RS232-C link. However a different method is commonly applied with low-cost sytsems when a CRT display is required, e.g. home computers which use domestic televisions for display purposes. In this case special function chips are available to extract picture information directly from memory and to convert this binary information into a video signal.

Figure 12.12 shows how characters are constructed on a CRT screen using a dot matrix for each character (the same method of constructing characters is used by a matrix printer). The CRT picture is created by driving an electron beam across the phosphor screen in a sequence of horizontal scans, and 10 scans are normally required to construct a complete row of characters. A spacing row and column surround each character matrix to prevent characters from merging together. The video waveform diagram shows how a pulse is used to modulate the video signal that is applied to the CRT in order to illuminate one dot in a matix. Typically 80 characters can be displayed on up to 64 rows, although normally only 40 (or even 24) rows are applied.

The circuit arrangement to generate this video signal is based on a CRT controller chip, as shown in Fig. 12.13. An area of microcomputer main memory RAM is reserved to hold data bytes which constitute the video information—1 byte for each character for normal text information. This information is extracted one byte at a time under DMA (Direct Memory Access) control, i.e. not under CPU and program control—see section 10.5. The CRT controller chip requests DMA transfer from the DMA controller

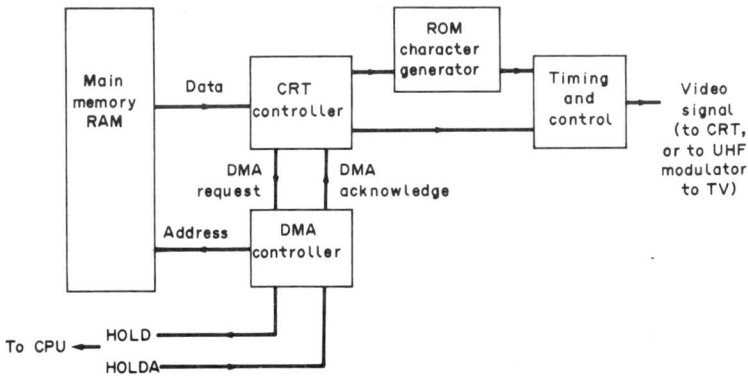

FIG. 12.13. Video generator circuit.

chip, which in turn generates a HOLD signal to the CPU in order to obtain use of the CPU's buses. A memory location is addressed by the DMA controller, and a data byte is read into the CRT controller. Typically a full row of 80 character bytes is extracted together and staticised within the CRT controller. Each character in turn is presented to the character generator ROM, which generates a group of bits to form one row in the dot matrix for that line of characters; this process is repeated for the other six rows for that line of characters. Whilst the seven rows of bits are being generated successively to create the video waveform for one row of 80 characters, the next group of 80 characters is extracted from microcomputer RAM (under DMA control).

If the video signal is to be applied to a domestic television receiver, it firstly passes through a UHF modulator (a small circuit encapsulated in a metal box). The output signal can then be connected directly to the aerial socket of the television receiver.

A similar arrangement of DMA transfer is used when the picture information is in the form of colour graphics in place of monochrome text. However more microcomputer RAM is required and the display information is stored in a different form, e.g. 1 bit represents a dot illumination and a further 3 bits indicate a code for the colour of that dot.

12.6 EXAMPLE INPUT/OUTPUT SYSTEM

A typical circuit arrangement that includes a PIO, UART and CTC is shown in Fig. 12.14. The Zilog PIO is used to drive a 4-digit segment display unit. Port A supplies the 8 segment signals which are multiplexed within the display unit to drive each of the 4 digit displays. The particular digit display which is selected to update with the segment pattern on Port A is determined by the setting of the 4 Digit lines—these are generated by one-half of Port B. Therefore drive software must set one Digit line together with the required Segment pattern for each of the 4 digit displays in turn. In this way each of the digit displays will be illuminated for one-quarter of the total time, i.e. one-quarter as brightly as for one digit displayed permanently. The remaining 4 output signals from port B drive a stepper motor (2 signals—Step and Direction), a loudspeaker and a single LED indicator.

The Intel UART connects to a VDU. The Zilog CTC is used to generate the clock signal for the UART (connected to TxC and RxC).

Notice that there is one spare chip select signal from the 2 to 4 decoder. This could be used if a second PIO is added to this circuit arrangement. Perhaps one port of this PIO could input 8 contact-closure signals (manual

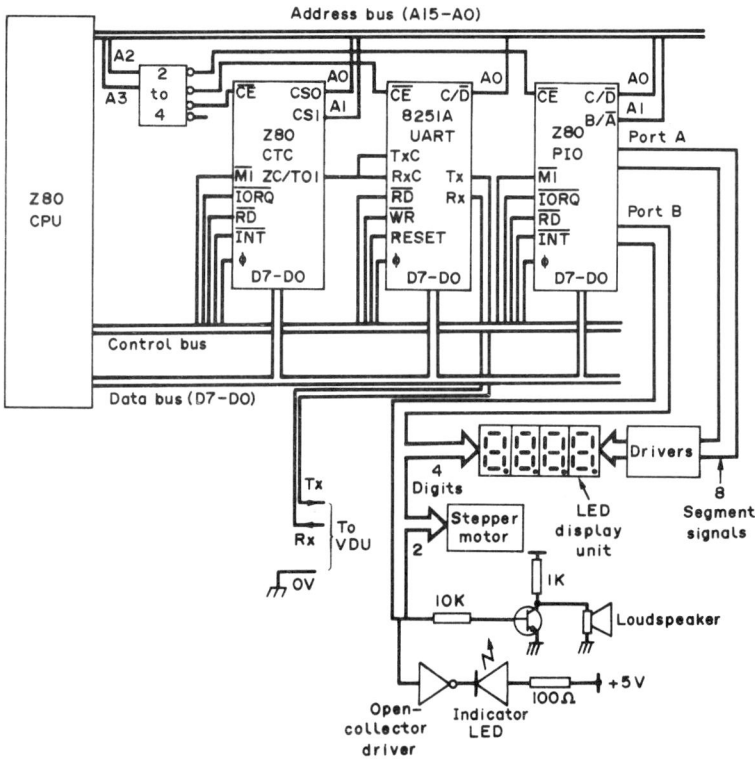

FIG. 12.14. Typical microcomputer input/output circuit.

keys, relay contacts, limit switches, etc.), and the other port could output 8 bits to a D/A converter chip to create an analogue signal (e.g. setpoint to a servo—position-control system).

The Zilog Z80 software that is needed to initialise the three programmable chips in Fig. 12.14 is as follows:

(a) *Initialising the PIO*

The PIO has 4 addresses:

$$Hex$$

Port A data   —00

Port A control—01   Notice that the PIO is selected ($\overline{CE} = 0$)

Port B data   —02   when A2 = 0 and A3 = 0

Port B control—03

Ports A and B are initialised to be output ports by sending hexadecimal 0F to each control register (4F sets a port to an input port), as follows:

```
LD   A,0FH      ;Control word for output port initialisation
OUT (01H),A     ;Send 0F to Port A control register
OUT (03H),A     ;Send 0F to Port B control register
```

This section of program normally resides at the beginning of the main program which is entered on start-up (e.g. power-up or reset), and it must be obeyed before the ports are used within the program.

(b) *Initialising the CTC*

The CTC has 4 addresses, one for each of the counter/timer circuits, and these addresses are hex. 08, 09, 0A and 0B. The Count Complete (see section 12.4) signal for the first of these circuits only is used in this application; its legend is ZC/T01. This counter is initialised to produce 1200 pulses/second (the counter is loaded with decimal 104 and it is set to decrement on every 16 input clock pulses ϕ, which are assumed to run at 2 MHz), as follows:

```
LD   A,05H      ;Control word (selects divide φ by 16), count to follow
OUT (08H),A     ;Send 05 to control register for first counter
LD   A,104      ;Count for 1200 pulses/second
OUT (08H),A     ;Send 104 to load first counter
```

For a full description of the options available on the Zilog CTC the reader is referred to the bibliography at the end of the chapter.

(c) *Initialising the UART*

Only one address line (A0) is connected to the UART, which effectively gives two addresses on the chip. However four addressable devices are obtained as follows:

	Hex	
Transmit (Tx)	—04	($\overline{WR}$ = 0 for an output instruction)
Receive (Rx)	—04	($\overline{RD}$ = 0 for an input instruction)
Control register	—05	($\overline{WR}$ = 0 for an output instruction)
Status register	—05	($\overline{RD}$ = 0 for an input instruction)

The control register is used to initialise the device in order to select the baud rate, number of data bits, parity and number of stop bits (see section 12.3). The status register is used to check if the last character has been cleared, if a parity error occurred on a received character or if the previous character had not been read by the CPU before a further character was received. The following section of program initialises the UART to run at 1200 baud, with 7 data bits, even parity and 1 stop bit:

```
LD   A,79H      ;Control word for baud rate, etc.
OUT (05H),A     ;Send 79 to control register
```

The output of two characters, e.g. letters P and T, can then be performed as follows:

```
        LD   A,50H      ;ASCII character for P
        OUT (04H),A     ;Output P to Tx
POLL:IN      A,(05H)    ;Input from status register
        AND 1           ;Mask out all bits except "Tx Ready" bit
        JP   Z,POLL     ;Repeatedly poll until character is cleared
        LD   A,54H      ;ASCII character for T
        OUT (04H),A     ;Output T to Tx
```

The reader is referred to the following bibliography for a full description of the initialising procedure for the 8251A UART.

BIBLIOGRAPHY

1. *Microcomputers and their Interfacing*, R. C. Holland, Pergamon, 1984.
2. *Elements of Microcomputer Interfacing*, Joseph J. Carr, Reston, 1984.
3. *Z80 Microprocessor Programming and Interfacing—Books 1 and 2*, Elizabeth A. Nichols, Joseph C. Nichols and Peter R. Rony, Prentice/Hall, 1979.

EXERCISES

1. What is "memory mapped input/output"? State an advantage and a disadvantage of its use.
2. What is meant by "programming" a PIO?
3. Is it possible to package a two-port PIO in a 24-pin DIL?
4. What is the primary advantage of parallel data transfer via an input/ output port to a VDU compared with serial transfer via a UART? What is the primary disadvantage if the VDU is 1 km from the computer?
5. Which characteristics of the RS232-C specification must be "initialised" into a UART before data transfers occur?
6. How many characters per second can be sent over a serial data link between two computers at 9600 baud (1 stop bit is used)?
7. Describe how a CTC can be used to generate a fixed time delay in a program.

8. Describe two different ways, one using a CTC, in which a micro-computer can count the number of pulses generated by a measuring instrument.

9. What is the principal advantage of applying a keyboard encoder chip in place of a PIO to read the operation of a manual keyboard?

10. What is the storage capacity of a single-density double-sided floppy disk which possesses 32 tracks, with each track divided into 20 sectors?

11. Where would you expect to find a shift register in the drive circuitry for a floppy disk?

12. Name the two microprocessor control bus signals that are applied in a handshaking mode when DMA transfers are applied.

13. Examine Fig. 12.14 and write a two-instruction program sequence that illuminates the indicator LED. Which item of input/output equipment would you discard so that the signal connections can be used for an inter-computer duplex (bi-directional) link?

CHAPTER 13

Typical Microprocessor Circuits and their Programming

13.1 PROGRAMMING LANGUAGES

The concept of "machine code" programming was introduced in chapter 10. In this technique program instructions are listed in hexadecimal byte form in the manner in which they are loaded into microcomputer memory (ROM or RAM) and then extracted and executed by the CPU. Assembly language programming allows the programmer to write his program using mnemonics, which are groups of letters that represent the machine code for each type of instruction. Assembly language programming is much easier than machine code programming and it eliminates the error-prone procedure of writing out the hexadecimal form of every instruction before the program is entered into memory. However it does require that an "assembler" program already exists within the machine's memory, and the assembler converts the mnemonic version of each instruction into machine code before entering bytes into memory.

Machine code programming and assembly language programming are termed "low level" languages. A programmer normally writes a low level language program in assembly language firstly before converting it into machine code if the machine does not possess an assembler. This is because it is easier to read and understand a program written using mnemonics and labels (names given to memory addresses)— example programs are given in section 10.6. In this case of machine code programming the microcomputer must possess a program in memory that allows the operator to enter the machine code program byte-by-byte—this program is commonly called the "monitor".

The alternative to a low level language is a high level language, in which program functions are written using commands which are much closer to spoken language. A short program written in a high level language converts into a much longer program in machine code. The advantages of high level language programming are that it is easier and quicker to write a program and the program can be transferred to a different computer with little or no modification, even though that computer may possess a different micro-processor with different machine code instructions. The most common high level language that is used with microcomputers is BASIC, although other languages that are encountered are PASCAL, C, FORTRAN, FORTH and several others. A program must exist within the microcomputer memory to convert the high level language program into machine code. This program is one of the following types:

(a) *interpreter*—this converts each line of the high level language program into machine code (often several instructions) and then executes that section for each line in the program every time the program is run.

(b) *compiler*—this creates a machine code file for the entire high level language program before the program is run; therefore there are two versions of the program, and the machine code version in memory is obeyed when the program is run.

A compiled version clearly runs much faster than a high level language program that is run via an interpreter. Home computers use interpreters, but commercial machines often use compilers.

The program examples given in this chapter are presented in low level language. This is because they perform detailed input/output functions, which are more easily specified in low level language form.

13.2   TRAINING MICROCOMPUTER HARDWARE ARRANGEMENT

Consider the hardware arrangement of a typical microcomputer training board, e.g. the Multitech Micro-Professor, shown in Fig. 13.1. A 4 K byte ROM (or EPROM) contains the "Monitor" program. This is the program that is automatically entered at switch-on and that obeys commands which are entered via the keyboard. These commands are to be enter bytes into memory, i.e. load a machine code program into the 2 K byte RAM, run that program and display the contents of registers and memory locations.

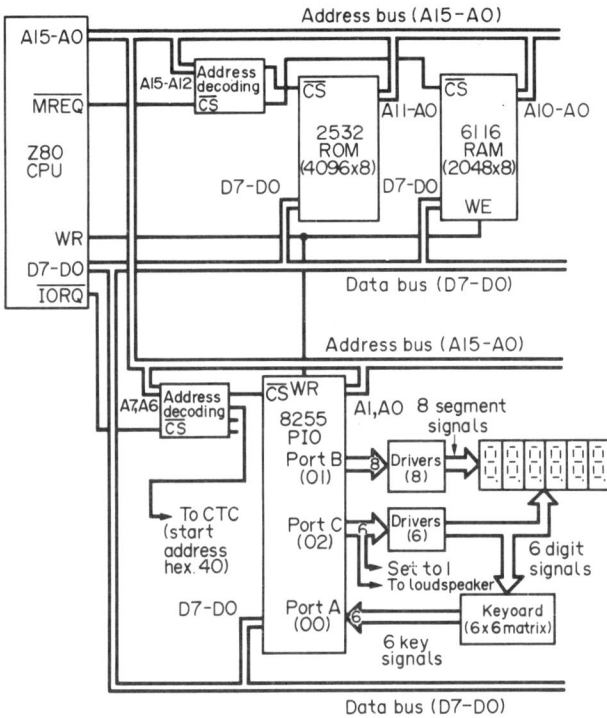

FIG. 13.1. Typical microcomputer training board (Multitech Micro-Professor).

The input/output circuit is driven by a single PIO, which supports three ports. Port B (address hexadecimal 01) drives the segment signals to a 6-digit LED segment display unit. Port C selects which of the 6 display digits is to respond to the segment pattern. Additionally Port C is used to select a group of 6 key settings which are to be "strobed in" to Port A from a keyboard—refer to section 12.5(a) for a description of the matrixing method used to connect several columns of keys onto a shared group of input lines. Notice that the last two bits on Port C are connected as follows:

(a) bit 6—must be set permanently to 1,
(b) bit 7—connected to a loudspeaker (via a transistor drive circuit).

Programming examples, which utilise this hardware configuration, are presented in the following sections. The RAM area, into which these programs can be loaded, commences at memory location hexadecimal 1800.

```
                    ┌─────────┐
                    │  Start  │
                    └─────────┘
                         │
    ┌────────────────────────────────────────┐
    │  ┌──────────────────────────────────┐  │
    │  │  Load accumulator with I         │  │
    │  └──────────────────────────────────┘  │
    │   ╱ Output accumulator to Port C ╱      │
    │  ┌──────────────────────────────────┐  │
    │  │ Load B register with delay count │  │
    │  └──────────────────────────────────┘  │
    │  ┌──────────────────────────────────┐  │
    │  │   Decrement B register           │  │
    │  └──────────────────────────────────┘  │
    │            ╱  B register > 0 ? ╲        │
    │   Yes     ╲                    ╱        │
    │            │ No                         │
    │  ┌──────────────────────────────────┐  │
    │  │  Load accumulator with O         │  │
    │  └──────────────────────────────────┘  │
    │   ╱ Output accumulator to Port C ╱      │
    │  ┌──────────────────────────────────┐  │
    │  │ Load B register with delay count │  │
    │  └──────────────────────────────────┘  │
    │  ┌──────────────────────────────────┐  │
    │  │   Decrement B register           │  │
    │  └──────────────────────────────────┘  │
    │            ╱  B register > 0 ? ╲        │
    │   Yes     ╲                    ╱        │
    └────────────────────│ No───────────────┘
```

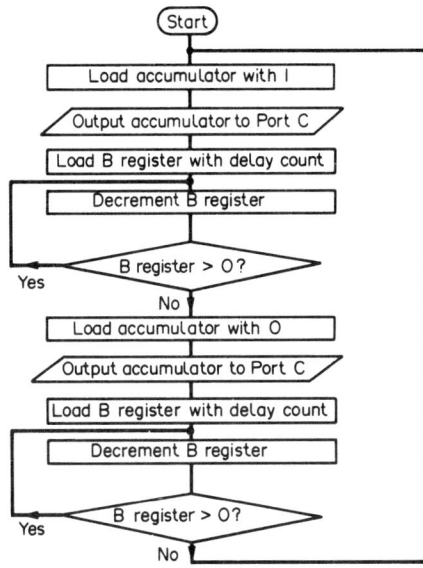

(a) Program flow chart

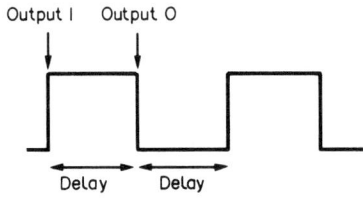

Output I   Output O

Delay   Delay

(b) Waveform produced on Port signal line to loudspeaker

FIG. 13.2. Program to generate a note.

13.3 PROGRAM TO GENERATE A NOTE

Figure 13.2 presents a "flow chart" of the short program which generates a continuous note (or tone) on the loudspeaker. The loudspeaker is caused to vibrate in order to generate the tone by sending a logic 1, followed by a delay, sending a logic 0, followed by another delay. If a capacitor in the drive transistor circuit feeding the loudspeaker rounds the sharp edges of the squarewave signal produced, then for a delay loop time of 0.5 ms a note of frequency

$$f = \frac{1}{T} = \frac{1}{2 \times \text{delay}} = \frac{1}{2 \times 0.5 \text{ ms}} = 1000 \text{ Hz}$$

is produced.

The shapes used in the flow chart for different processing operations are:

oval—for start and end (these are flow chart delimiters only, i.e. they do not translate into actual program instructions); there is no end block in this program because it continually repeats itself;

rectangle—for normal processing stages;

parallelogram—for input and output operations;

diamond—for making decisions (there are always two output paths).

This flow chart translates into the following program:

Object Program		Source Program			
Memory address	Machine code	Label	Mnemonic	Operand	Comments
			ORG	1800H	Load program at hex. 1800
1800	3E,C0	START	LD	A,C0H	Load A with 1100 0000
1802	D3,02		OUT	(02H),A	Output A to Port C
1804	06,32		LD	B,50	Load B with delay loop count of 50
1806	05	→LOOP1	DEC	B	Decrement B
1807	C2,06,18		JP	NZ,LOOP1	Jump if B is not zero
180A	3E,40		LD	A,40H	Load A with 0100 0000
180C	D3,02		OUT	(02H),A	Output A to Port C
180E	06,32		LD	B,50	Load B with delay loop count of 50
1810	05	→LOOP2	DEC	B	Decrement B
1811	C2,10,18		JP	NZ,LOOP2	Jump if B is not zero
1814	C3,00,18		JP	START	Jump to beginning of program
			END		End of program

The left-hand bit in the accumulator (A register) sets the bottom bit on Port C. Notice how the next bit must always be set to 1 due to external hardware requirements (undefined). The program delay is created by simply loading a register with a number and decrementing it to zero in a delay loop. When the second delay loop is completed one squarewave is generated. The last instruction in the program is an unconditional jump, such that the program loops repeatedly in order to generate a continuous tone.

Notice that the same program can be used to rotate a stepper motor if the port address is changed.

13.4 PROGRAM TO GENERATE A TUNE

The flow chart of a program to generate a tune is shown in Fig. 13.3. The program uses two data lists (or "data tables") in memory—these hold the notes (delays to create notes of specific frequencies) and their durations (numbers of square waves). The program listing, in assembly language only, is as follows:

```
          ORG 1800H        ;Load main program at 1800
   START:LD   E,18         ;Loop count (18 notes in tune)
          LD   IX,1900H    ;Set IX to address of frequency data table
          LD   IY,1940H    ;Set IY to address of duration data table
→TUNE:LD   B,(IX)          ;Load B with frequency (delay for
                                squarewave)
          LD   C,(IY)       :Load C with duration (number of
                                squarewaves)
→NOTE:LD   A,C0H           ;Output 1
          OUT (02H),A      ;          to loudspeaker
          CALL DELAY       ;Delay on B (call subroutine)
          LD   A,40H       ;Output 0
          OUT (02H),A      ;          to loudspeaker
          CALL DELAY       ;Delay on B (call subroutine)
          DEC C            :Decrement duration count
─────JR   NZ,NOTE          ;Repeat (generate 1 note)
          INC  IX          ;Increment pointers
          INC  IY          ;                 for data tables
          DEC  E           ;Decrement note count
─────JR   NZ,TUNE          ;Repeat (generate 18 notes)
          HALT             ;Stop program on this instruction (tune
                                complete)
                           ;Main program complete, load
                                subroutine
          ORG 1850H        ;Load subroutine at 1850
   DELAY:LD   D,B          ;Transfer frequency (delay) to D
→PAUSE:DEC D               ;Delay
─────JR   NZ,PAUSE         ;      on D
          RET              ;Return to main program
          END              ;End of program
```

IX and IY registers are used in indirect addressing mode to access values from the two data tables. The program section that creates a time delay in the construction of the squarewave is loaded as a subroutine, and called twice by the main program. The inner loop, which includes the jump to

Main program

```
                        ( Start )
                           │
┌──────────────────────────────────────────────────┐
│         Load E with number of notes               │
├──────────────────────────────────────────────────┤
│ Load IX with memory address of frequency data table│
├──────────────────────────────────────────────────┤
│ Load IY with memory address of duration data table │
└──────────────────────────────────────────────────┘
      ┌────────────────────────────────────────────┐
      │   Load B with frequency (using IX indirect addressing) │
      ├────────────────────────────────────────────┤
      │  Load C with duration (using IY indirect addressing) │
      ├────────────────────────────────────────────┤
      │   Output I to loudspeaker (2 instructions)  │
      ├────────────────────────────────────────────┤
      │  Call DELAY subroutine (delays on freqency in B) │
      ├────────────────────────────────────────────┤
      │   Output O to loudspeaker (2 instrutions)   │
      ├────────────────────────────────────────────┤
      │  Call DELAY subroutine (delays on frequency in B) │
      ├────────────────────────────────────────────┤
      │      Decrement duration count in C          │
      └────────────────────────────────────────────┘
  Yes          Duration count > O?
      ──────────────────────────────
                   No
      ┌────────────────────────────┐
      │     Increment IX and IY     │
      ├────────────────────────────┤
      │   Decrement note count in E │
      └────────────────────────────┘
  Yes          Note count > O?
      ──────────────────────────────
                   No
      ┌────────────────────────────┐
      │           Halt              │
      └────────────────────────────┘
                 ( End )
```

Subroutine

```
                ( DELAY )
                    │
      ┌────────────────────────────────┐
      │  Load B into D (delay count)    │
      ├────────────────────────────────┤
      │   Decrement delay count in D    │
      └────────────────────────────────┘
  Yes        Delay count > O?
      ──────────────────────────────
                  No
      ┌────────────────────────────────┐
      │     Return to main program      │
      └────────────────────────────────┘
```

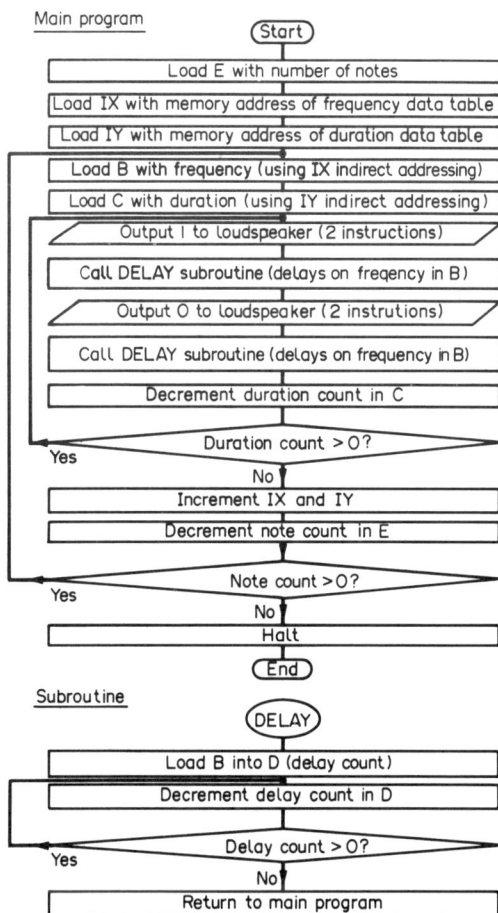

(a) Program flow chart

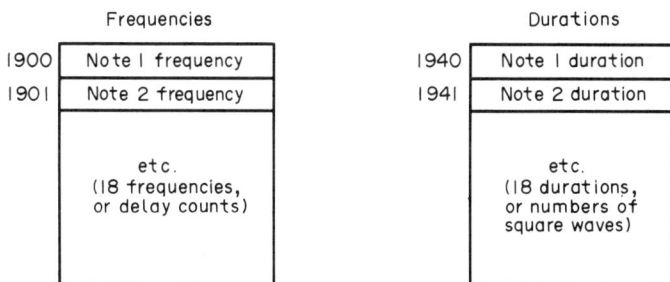

Frequencies		Durations	
1900	Note 1 frequency	1940	Note 1 duration
1901	Note 2 frequency	1941	Note 2 duration
	etc. (18 frequencies, or delay counts)		etc. (18 durations, or numbers of square waves)

(b) Data tables in memory

FIG. 13.3. Program to generate a tune.

NOTE instruction, generates a single note. The outer loop, which includes the jump to TUNE instruction, generates a series of notes, i.e. a complete tune. A musical reader can calculate his own values for the two data tables for a tune of his choice if he assumes that the time taken for one pass through the two-instruction delay loop (at PAUSE) is 10 μs, and that:

$$\text{middle c} = 960 \text{ Hz}$$
$$\text{and C, 1 octave higher} = 1920 \text{ Hz}$$

The tune can be made to repeat continuously if the HALT instruction is replaced with: JR   START

Notice that relative jump instructions are used in this program in place of absolute jump instructions (JR in place of JP) because they are shorter—1 byte shorter in machine code for each jump instruction.

13.5 PROGRAM TO DRIVE A SEGMENT DISPLAY

The method of software driving a multi-digit segment display unit is to illuminate each digit in turn. Although only one digit is illuminated at any time, the eye is deceived into believing that all digits are illuminated equally if the update speed is fast enough. The intensity of illumination for the six digit displays in Fig. 13.1 is one-sixth of that which is achieved when only a single digit is displayed permanently.

The program to display six digits, or characters, on the display unit of Fig. 13.1 outputs a segment pattern on Port B, together with the output of a single bit on Port C (to select the particular display digit which is to receive that segment pattern), for each digit in turn, as follows:

```
          ORG  1800H        ;Load main program at 1800
   START:LD    B,6          ;Loop count (6 digits)
          LD    HL,1900H    ;Set HL to data table start address
→ LOOP:LD     A,(HL)       ;Load A from data table (segment byte)
          OUT  (01H),A      ;Output A to segment port
          INC  HL           ;Increment HL (data table pointer)
          LD    A,(HL)      ;Load A from data table (digit byte)
          OUT  (02H),A      ;Output A to digit port
          INC  HL           ;Increment HL (data table pointer)
          CALL DELAY        ;Delay (display character for a short
                            ;   period)
          LD    A,40H       ;Blank
          OUT  (02H),A      ;       out digitis
          DEC  B            :Decrement loop count
          JR    NZ,LOOP     ;Repeat for each character
```

```
        JP    START      ;Repeat whole program
                         ;Main program complete, load
                             subroutine
        ORG  1860H       ;Load subroutine at 1860
DELAY:LD    C,FFH        ;Delay count
→PAUSE:DEC  C            ;Delay (255 × 10 μs
└───────JP    NZ,PAUSE       = 2.5 ms approximately)
        RET              ;Return to main program
        END              ;End of program
```

Notice that a delay is inserted into the program to cause each digit to be displayed for 2.5 ms before the next digit is updated. Without this delay the slow response time of the LED segments causes low illumination levels. The output operation to blank all digits momentarily after each digit has been displayed is to prevent the following segment pattern appearing on the same digit position for a short period before the correct digit setting is output on Port C. The data which the program requires is assembled in a data table commencing at memory location 1900, as shown in Fig. 13.4. If the segments bits are connected to the display unit in the following order on Port B:

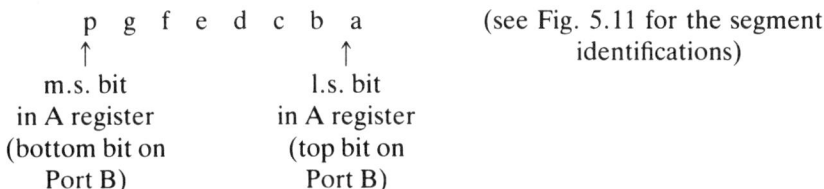

```
  p   g   f   e   d   c   b   a        (see Fig. 5.11 for the segment
  ↑                       ↑                     identifications)
m.s. bit                l.s. bit
in A register         in A register
(bottom bit on        (top bit on
  Port B)               Port B)
```

then a bit pattern of 0100 1111 (hex. 4F) must be loaded into memory location 1900 if it is required to display the number 3 on the right-hand digit display (assuming that the top bit on Port C selects the right-hand digit).

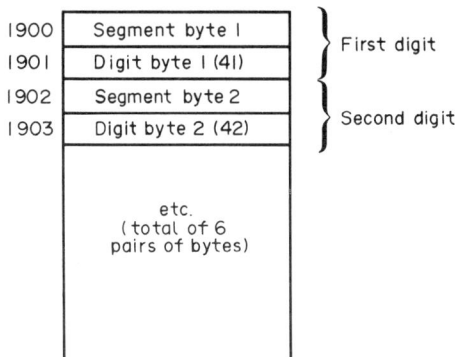

1900	Segment byte I
1901	Digit byte I (41)
1902	Segment byte 2
1903	Digit byte 2 (42)
	etc. (total of 6 pairs of bytes)

First digit } (1900, 1901)
Second digit } (1902, 1903)

Fig. 13.4. Data table of segment display program.

13.6 PROGRAM TO DISPLAY TIME-OF-DAY

In this final program example the hardware configuration of Fig. 13.1 is used to display the time-of-day on the segment display unit. Two programs are applied. The first (the main program) displays six numbers representing the current time—these numbers are held in six consecutive memory locations. The other program, which is interrupt-driven by the counter/timer chip (CTC), updates these six numbers every second. The main program consists of two halves:

(a) initialise the CTC (to provide 218 interrupt pulses every second) and set the interrupt vector (start address of interrupt program)

(b) display the six time-of-day counts in a continuous loop

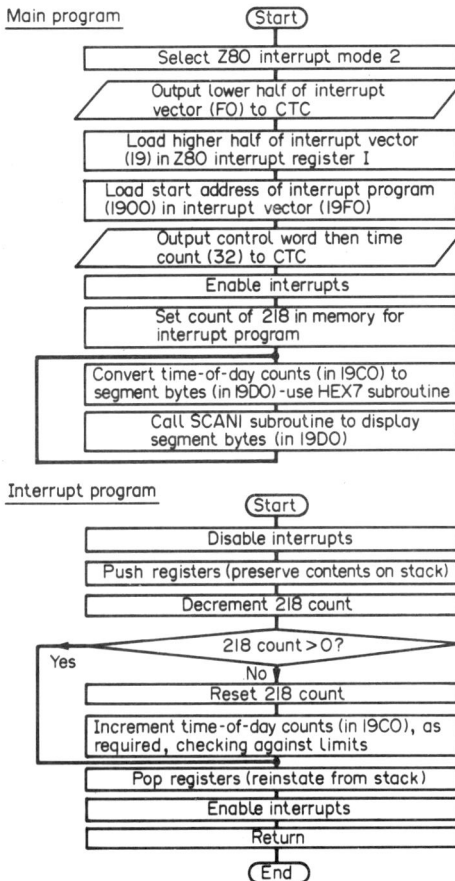

Main program   (Start)

| Select Z80 interrupt mode 2 |

| Output lower half of interrupt vector (FO) to CTC |

| Load higher half of interrupt vector (19) in Z80 interrupt register I |

| Load start address of interrupt program (1900) in interrupt vector (19FO) |

| Output control word then time count (32) to CTC |

| Enable interrupts |

| Set count of 218 in memory for interrupt program |

| Convert time-of-day counts (in 19CO) to segment bytes (in 19DO)-use HEX7 subroutine |

| Call SCANI subroutine to display segment bytes (in 19DO) |

Interrupt program   (Start)

| Disable interrupts |

| Push registers (preserve contents on stack) |

| Decrement 218 count |

218 count > 0?   Yes

No

| Reset 218 count |

| Increment time-of-day counts (in 19CO), as required, checking against limits |

| Pop registers (reinstate from stack) |

| Enable interrupts |

| Return |

(End)

FIG. 13.5. Flow charts for time-of-day programs.

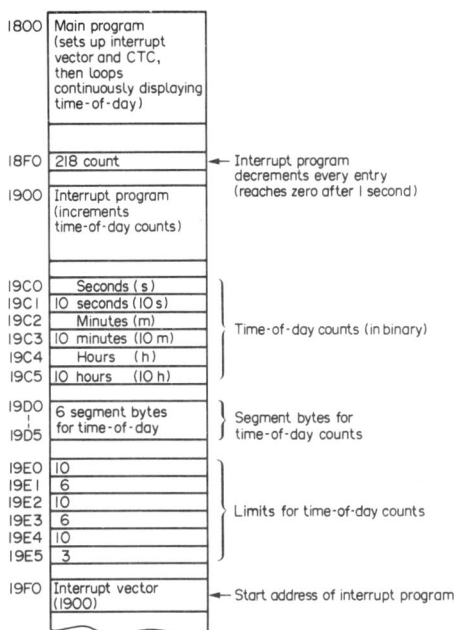

FIG. 13.6. Memory map of time-of-day program system.

Therefore when the main program is entered it performs stage (a) above once, and then continually loops through stage (b) providing a permanent display. The interrupt program, which is called 218 times a second, updates the six counts in memory as required every second. It uses a memory location (hex. 18F0) to hold a decrementing count of the number of times it is entered in order that it can detect when 1 second has elapsed. The first of the four channels on the Zilog CTC—see sections 12.4 and 12.6(b)—is used, and the count of 218 is obtained as follows:

Clock pulse (CPU clock) repetition rate $= 1.7898$ MHz

$$\text{Clock pulse time} = \frac{1}{1.7898 \text{ MHz}} = 0.55872 \ \mu s$$

If the CTC is initialised to decrement the counter every 256 clock pulses, and the counter is loaded initially with 32:

Interrupt program is entered every interrupt pulse
$= 32 \times 256 \times 0.55872 \ \mu s = 4.577$ ms

i.e. 218 times/second

The flow charts for the two programs are shown in Fig. 13.5, the memory map is shown in Fig. 13.6 and the program listing is as follows:

Main program

```
              ORG 1800H              ;Load main program at 1800
              IM    2                ;Set interrupt mode 2
              LD    A,F0H            ;Output half interrupt
              OUT (40H),A            ;                      vector (F0)
                                     ;                         to CTC
              LD    A,19H            ;Set other half interrupt vector (19)
              LD    I,A              ;            in Z80 interrupt
                                     ;                      register I
              LD    A,00H            ;Set 1900 (start address
              LD    (19F0H),A        ;        of interrupt service
              LD    A,19H            ;                routine) in interrupt
              LD    (19F1H),A        ;                      vector (19F0)
              LD    A,B5H            ;Output control
              OUT (40H),A                           word to CTC
              LD    A,32             ;Output time count
              OUT (40H),A            ;                 to CTC
              EI                     ;Enable interrupts
              LD    A,218            ;Store 218 (for 1 second count)
              LD    (18F0H),A        ;     in 18F0 for interrupt program
REPEAT:LD     B,6                    ;Loop count of 6 (6 time counts)
              LD    IX,19C0H         ;Address of time-of-day counts
              LD    IY,19D0H         ;Address of segment bytes
CONVERT:LD    A,(IX)                 ;Fetch one time-of-day count
              CALL HEX7              ;Convert to segment byte (use
                                                  subroutine HEX7)
              LD    (IY),A           ;Store in segment byte data table
              INC IX                 ;Increment data
              INC IY                 ;                      table pointers
              DEC B                  ;Repeat locp for
              JP    NZ,CONVERT       ;            6 time counts
              LD    IX,19D0H         ;Set IX for SCAN1 subroutine
              CALL SCAN1             ;Display 6 counts (use
                                                  subroutine SCAN1)
              JP    REPEAT           ;Continually update display
```

Interrupt program

```
              ORG 1900H              ;Load interrupt program at 1900
              DI                     ;Disable interrupts
              PUSH AF                ;Store A and F (flags) on stack
              PUSH BC                ;Store B and C on stack
```

```
        PUSH IX          ;Store IX on stack
        PUSH IY          ;Store IY in stack
        LD   A,(18F0H)   ;Load A with 218 count
        DEC  A           ;Decrement 218 count
        LD   (18F0H),A   ;Put back into memory
        JR   NZ,EXIT     ;Jump if not zero (has 1 second elapsed?)
        LD   A,218       ;Reload 218
        LD   (18F0H),A   ;        count in memory
        LD   IX,19C0H    ;Set IX to address of time-of-day counts
        LD   IY,19E0H    ;Set IY to address of limits of times
        LD   B,6         ;Loop count of 6—check 6 times
CHECK:LD   A,(IX)      ;Pick up 1 of 6 counts (s, 10 s, m, 10 m, h, 10 h)
        INC  A           ;Increment time count
        CP   (IY)        ;Compare with limit
        LD   (IX),A      ;Put back into memory
        JR   NZ,HOME     ;Jump if not same as limit (less than limit)
        LD   A,0         ;Load zero
        LD   (IX),A      ;        into time count
        INC  IX          ;Increment IX (next time count)
        INC  IY          ;Increment IY (next limit)
        DEC  B           ;Repeat loop
        JR   NZ,CHECK;            6 times
HOME:LD   A,(19C5H)   ;Check for midnight—pick up 10 h
        CP   2           ;Compare with 2
        JR   NZ,EXIT     ;Jump if 0 or 1 (less than 20 hours)
        LD   A,(19C4H)   ;Pick up h
        CP   4           ;Compare with 4
        JR   NZ,EXIT     ;Jump if 0, 1, 2 or 3 (less than 23 hours)
        LD   A,0         ;Midnight—0 in A
        LD   (19C4H),A   ;Store 0 in h
        LD   (19C5H),A   ;Store 0 in 10 h—other time counts already 0
EXIT:POP  IY          ;Reinstate IY from stack
        POP  IX          ;Reinstate IX from stack
        POP  BC          ;Reinstate B and C from stack
        POP  AF          ;Reinstate A and F (flags) from stack
        EI               ;Enable interrupts
        RETI             ;Return to main program
```

The first part of the main program (as far as EI instruction) sets up the interrupt vector (half sent to CTC, half loaded into CPU) and initialises the CTC—refer to the bibliography (3) for detailed descriptions of these procedures. The second half of the main program performs a continuous looping operation in order continually to update the segment display with the six time counts representing the time-of-day. A subroutine SCAN1 is used to display the six numbers. This subroutine resides in the Monitor ROM, and it requires that IX shall be loaded with the start address (19D0) in memory of the six numbers to be displayed. Also it requires that these six numbers shall be stored in segment pattern form and not in binary. Therefore the inner loop (around CONVERT) converts the six binary numbers (commencing at 19C0) into segment pattern form using a second subroutine HEX7. HEX7 converts the binary number in A register into a segment pattern byte and places that byte back into A register.

218 times every second the main program is interrupted. The interrupt program at its commencement preserves the contents of the registers which it uses; it reinstates them at the end of the program. It firstly checks if it has decremented the 218 count to zero. If it has not, then it exits and returns to the main program. If the 218 count is zero, i.e. 1 second has elapsed, it resets the count back to 218 and enters a loop (around CHECK). This loop increments the seconds count and checks each of the six time counts in turn, resetting them to 0 when they exceed limits, i.e. 9 for seconds, 5 for tens of seconds, etc. These limits are held in a data table commencing at memory location 19E0. After the time has been updated in this way, the interrupt program checks for midnight (counts of 240000) and resets the hours and tens of hours to zero if the time is midnight. Program control is then returned to the main program.

This type of program arrangement is required for any computer that requires to keep (and perhaps display) an accurate time-of-day. If the interrupt program forms a module of a larger program in a multi-programming or multi-user system, then programs can be called with specified time delays by other programs or at specific times of the day.

BIBLIOGRAPHY

1. *An Introduction to Microelectronics and Microprocessor Systems*, G. H. Curtis and P. G. Wilks, Stanley Thomas, 1982.
2. *Microelectronic Systems 2 Checkbook*, R. E. Vears, Butterworth, 1982.
3. *Microcomputers and their Interfacing*, R. C. Holland, Pergamon, 1984.

EXERCISES

1. Describe the principal differences between a high level language interpreter and compiler.

2. What alteration would you make to the program in section 13.3 in order to double the frequency of the note produced?

3. Write a Z80 assembly language program section that generates a time delay of 0.5 s. (Hint: place a delay loop which decrements a register count inside another delay loop which decrements a different register count.)

4. Assume that indicator LEDs of different colours are connected to a Z80 microcomputer output port. Write an assembly language program that illuminates the LEDs in a sequence that is related to that applied with a traffic light system.

5. Write a Z80 assembly language program that scans an 8 × 8 matrix keyboard via a PIO, and illuminates a LED when a specific key is pressed.

6. Extend the program in 5 to sound a different note on a loudspeaker for operation of each keyboard key, i.e. generate an electronic organ.

7. Write a program that provides a display of the time between successive operations of a manual pushbutton, i.e. generate a microcomputer stop-watch. Use a CTC-driven timer interrupt to provide accurate time-keeping.

CHAPTER 14

Fault Finding

14.1 FAULT TYPES

Modern circuit boards for the whole range of electronic applications consist predominantly of ICs, supported by some discrete components, e.g. capacitors for removing noise across dc supply pins, transistors for high current signals. Faults that occur with circuit boards fall into one of the following categories:

(a) *Chip failure* (or failure of supporting discrete component, e.g. capacitor)

A chip (IC) failure can be due to failure of the internal logic or due to a short-circuit or open-circuit of the internal connections to the interconnecting pins. The precise nature of the fault matters little—the faulty chip must be replaced. The replacement of a chip which is mounted in an IC socket is clearly much more straightforward than the replacement of a chip which is soldered into the circuit board. Care must be taken when inserting an IC to ensure that it is not inserted backwards. The chip will not only fail to operate correctly but it may well be destroyed because the power supply to it is reversed and overheating may occur.

The MTBF (Mean Time Between Failures) of ICs is normally extremely high, so it is unwise to jump to conclusions and immediately replace suspect chips. Unless a chip is obviously overheated or damaged the fault probably lies elsewhere.

(b) *Open-circuit fault*

A circuit break is one of the most common sources of faults. It may be caused by a broken solder connection ("dry joint"), IC making poor contact in its socket (due to dirt, loose connection, bent pin), ruptured component wire (perhaps caused by repeated bending of component, e.g. capacitor), break in copper track or poor contact on edge-connector in a multi-board system. Visual inspection and physically tapping the board and its ICs may help to locate such faults.

(c) *Short-circuit fault* ("Bridge" fault)

Typical causes are adjacent pins in an IC are touching or the introduction of a foreign body makes contact between adjacent pins or copper tracks. Care must be taken to prevent strands of wire, solder blobs and even items of mechanical hardware (screws, wire-cutter, etc.) falling onto circuit boards.

(d) *Power supply failure*

This most obvious cause of system failure is often the last item to be tested. The dc supply voltage may be zero due to a fault in the power supply or due to a ruptured fuse, or it may be set at an incorrect level. A circuit fault may draw excessive current and pull the power supply voltage level low; in this case therefore it is sensible to disconnect the circuit board from the power supply to indicate the location of the fault (power supply or circuit board). If dc is set correctly on the circuit board it is sensible to test further for correct dc supply to each IC, especially if an IC feels cold some time after switch-on.

(e) *Electromechanical failure*

This particularly applies with microcomputers, when damaged floppy disks, broken keyboard keys or other problems associated with moving parts of electromechanical devices cause faults in peripheral equipment.

(f) *Miscellaneous*

Other occasional problems may arise as a result of external interference from sources of electromagnetic radiation, e.g. equipment that generates electrical sparks. Alternatively timing problems may occur in electronic systems which are running at the upper limit of speed tolerance of the circuit, e.g. the rise and fall times of fast pulses may not be adequate to cause the triggering of a circuit element.

In this chapter various items of test equipment, which can help to locate faults are described. However several preliminary checks can be made and can identify problems. For example visual inspection can locate some open-circuit and short-circuit faults on circuit boards. The failure of indicator LEDs and displays to operate may identify a fault area. Overheated and damaged components can be detected by their charred or blackened appearance. The sense of touch can help to locate some chip faults, e.g. a cold IC may not be receiving dc power, whilst a hot IC is a sign of probable permanent damage to the chip (but beware of severely over-heated devices).

Clearly one of the most important aids to any Test Technician when he is undertaking a fault-finding investigation is an understanding of the system he is working on. He should appreciate the overall role of the system and interpret correctly the manifestations of a problem. It is not unusual for non-technical operators of complicated electronic systems, e.g. computers, to report a fault with an inaccurate description—indeed there may be no fault at all. Abnormal conditions that preceded a failure should be identified, e.g. an interconnecting cable or plug/socket arrangement may have been disturbed or damaged. The Technician should read carefully the manufacturer's manual (perhaps test waveforms are given) and ensure that he understands the operation of the circuit before he makes test measurements. The correct operation of test equipment should be verified. The crucial effect of operating temperature should be appreciated. Many faults only become apparent some time after switch-on due to a heating problem; loss of cooling air from a cooling fan can cause overheating in a multi-board system. The application of a cooling spray or heat gun (blows hot air) can help to locate heat-dependent faults on ICs and other components. It can be helpful in a fault-finding exercise if an identical operational system is available, and waveforms at different test points in the circuit can be compared with those in the faulty system.

14.2 OSCILLOSCOPE AND MULTIMETER

The two traditional items of test equipment that are used to help locate faults in electronic systems are the oscilloscope (or CRO—Cathode Ray Oscilloscope) and the multimeter. The latter may be a conventional analogue indicator, i.e. swinging pointer driven by a moving coil, and the dominant product has been the "Avometer", or it may be a digital indicator, i.e. the DVM (Digital Voltmeter). The DVM possesses several advantages, e.g. it loads the circuit node under test much less because it has a higher input impedance for voltage measurement, it is more portable, read-out is more precise, it may possess self-calibration. The CRO is invaluable for indicating voltage waveforms.

Both devices can be applied to analogue and digital circuits. The items of test equipment described in later sections of this chapter are only applied to digital circuits (including microprocessor and computer systems).

An example of the use of a CRO and DVM is for the testing of the dc power supply to an electronic system, as shown in Fig. 14.1. If +5 V dc does not appear at the output connections of the power supply the voltage signals at various points in the power supply can be examined. The CRO timebase must be switched to a slow setting (e.g. 10 ms per division) to observe the ac signals. Notice that the CRO can also be used to measure dc levels, although

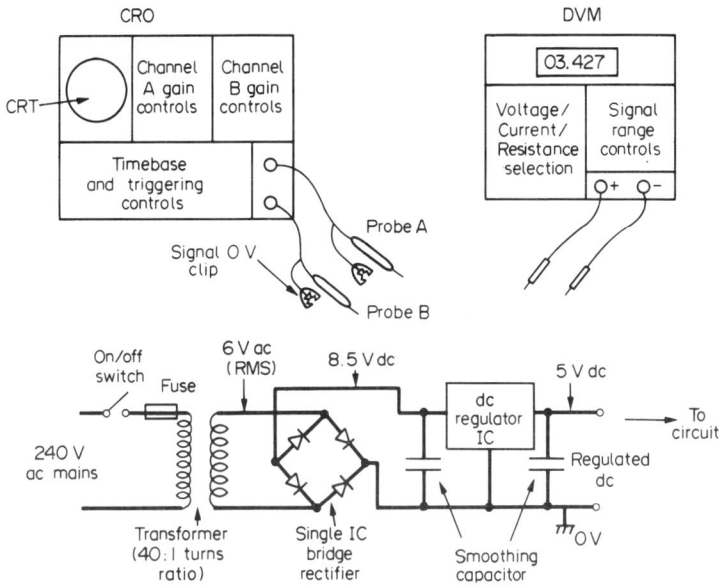

FIG. 14.1. Use of CRO and DVM to test dc power supply.

measurement accuracy is not as good as the DVM. The DVM may be useful to check current delivered to the electronic circuit and even circuit continuity (on resistance selection) of the fuse and transformer windings.

The same items of test equipment can be applied to measure test points around the circuit which is faulty, as follows:

(a) The DVM can be applied to measure dc voltages in an analogue system that processes instrumentation signals, whilst the CRO can measure ac voltage in an analogue system.

(b) Digital systems can be monitored by the DVM for dc supply voltages and also for logic levels that are static and can be predicted. The CRO can display pulse waveforms and dynamic effects in switching systems.

(c) The CRO can be used to detect noisy signals. For example coupling from one IC to another through the power supply due to rapid signal variations and therefore power supply load variations can occur, and the CRO can be used to indicate that the power supply voltage has a small voltage superimposed across it. This problem can be overcome by inserting a ceramic disc capacitor across the power supply pins on each IC as shown in Fig. 14.2. These fast-response capacitors augment the slow-response electrolytic capacitor that is connected across the supply lines in the power supply.

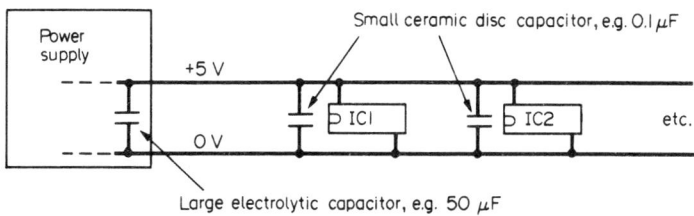

Fig. 14.2. Localised de-coupling capacitors in electronic circuit.

(d) The DVM can be applied in gating systems to check the current flow
at gate inputs if the IC under test is firstly disconnected from its
surrounding circuits. If the input currents on adjacent inputs to a gate
are not identical for the same logic levels then that IC is suspect.
Excessive output currents can also be detected.

Many other measurements can be made and many different systems
tested with a CRO and DVM of course. Whilst more sophisticated test
equipment, which will be described in later sections, has been developed
recently, the ubiquitous natures of the CRO and DVM guarantee them
essential roles in any fault-finding department or workshop.

14.3 LOGIC PROBE, LOGIC PULSER AND CURRENT TRACER

Three hand-held items of test equipment have evolved with the proliferation of digital circuits, including computer circuits. These small and relatively
cheap devices are:

(a) logic probe—indicates a logic level with a visual indicator (normally a
LED).
(b) logic pulser—generates pulses to inject into a circuit under test.
(c) current tracer—indicates by varying illumination of a LED the magnitude of current along a conductor.

Whilst a logic probe is often applied to test a suspect circuit on its own, it
is commonly used together with a logic pulser, as shown in Fig. 14.3. Each
device requires a dc power source, and normally this is "borrowed" from the
circuit under test by attaching the two clips to any convenient +5 V and 0 V
points. The indicator LED on the logic probe can be used to indicate logic
levels at any point, e.g. IC pin, around the circuit. Typically the LED glows
brightly for logic 1, does not illuminate at all for logic 0 and glows with a low
intensity for the "floating" state (electrically unconnected). Some logic
probes possess two LEDs of different colours to indicate the two different

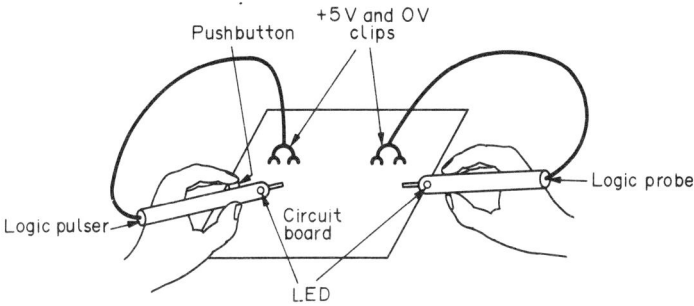

FIG. 14.3. Use of logic probe and logic pulser.

logic levels. Additionally some logic probes possess a selector switch to provide selection for testing either TTL or CMOS circuits.

A particularly useful feature with many logic probes is that of "pulse stretching", i.e. they slow down fast pulse waveforms so that the flicker rate of the indicating LED is slow enough to be detected visually. A pulse repetition rate of up to 30 MHz can be slowed to provide indication, whilst a single narrow pulse (e.g. 100 ns wide) can be "stretched" and displayed.

The logic pulser also possesses an indicating LED at its tip to indicate to the operator when pulses are being injected. Operation of the pushbutton causes pulses to be generated, and these pulses can be selected to be:

(a) continuous,
(b) burst of pulses,
(c) single pulse.

The logic probe can be used to detect the presence of these pulses at later points in the circuitry. For example, pulses can be injected at a gate output and monitored at the succeeding gate input to check for circuit continuity. Alternatively pulses can be injected at a gate input and correct gate operation can be confirmed by monitoring the gate output.

A variation of the logic probe is a logic monitor (or logic clip), which displays the state of several signals and is illustrated in Fig. 14.4. This

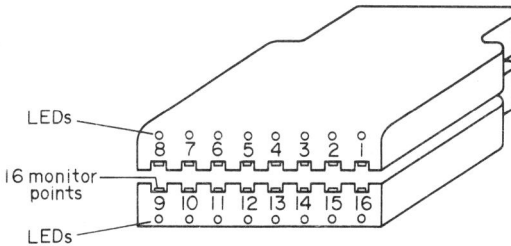

FIG. 14.4. Logic monitor.

particular logic monitor is designed for 16-pin DIL ICs, and it clips over the IC which is under test (this is called "piggybacking"). It derives its dc power from the IC, and displays the states of the various signals on its LEDs, e.g. LED is lit for logic 1, not lit for logic 0.

The final small hand-held item of test equipment which is available is the current tracer. This has a pencil-shape like the logic probe and logic pulser, but its tip does not have to be placed in contact with a test point—the tip contains a magnetic sensor that detects the field produced by the current flow in the circuit under test. An example of its use is shown in Fig. 14.5. One of the wire-AND (see section 3.4, Fig. 3.8) gates exhibits a fault due to an internal short-circuit which causes it to draw current when it should be non-conducting. The logic pulser injects pulses at the common output point and the current tracer can be used to trace these current pulses back into the faulty gate; the current tracer LED illuminates on the output of the faulty gate only, and not on the other gates. The sensitivity control must be adjusted before tests are made, and care must be taken to align the current tracer in the direction of the current flow. The device "stretches" pulses in the same manner as the logic probe so that pulsing of the indicator LED can be detected visually. Other fault types that can be investigated with a current tracer are:

(a) shorted copper tracks or IC pins,
(b) power supply short-circuits, e.g. $+V_{CC}$ to ground,
(c) open-circuits through copper tracks and through components, e.g. resistors,
(d) IC input shorted to ground.

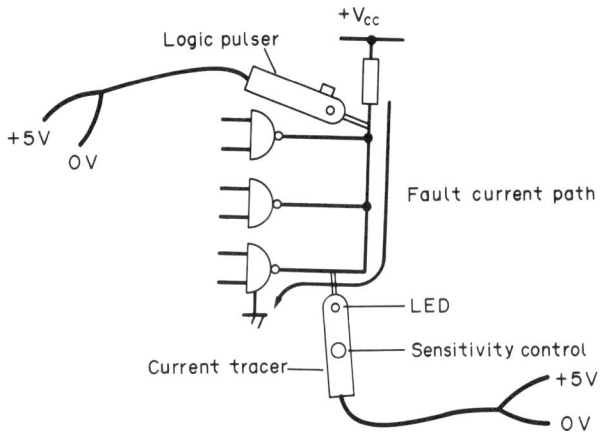

FIG. 14.5. Use of current tracer (on wire-AND circuit).

One practical tip that is worthy of mention here is that if tests with one of the above items of test equipment indicates that a chip possesses an open-circuit at one of its inputs or outputs, correct circuit operation can sometimes be achieved by simply piggybacking a good IC on top of the faulty IC.

14.4 LOGIC ANALYSER

A logic analyser is an item of digital test equipment that provides a wide range of data about a circuit under test. It monitors a large number of signals and employs a CRT to display information about these signals in a variety of forms. It is an expensive piece of equipment (costing many thousands of pounds), but is particularly useful for fault-finding in microprocessor and computer systems.

Figure 14.6 shows the application of a logic analyser to a circuit board that is microprocessor-based. The logic analyser, which looks much like an elaborate VDU, possesses 32 probes (some possess less) and in this case 16 are connected to the address bus lines and 8 are connected to the data bus lines. The logic analyser possesses its own internal memory and can staticise the setting of the various test signals after a chosen "trap" condition, i.e. a specified combination of 1s and 0s on a selected group of test signals. The settings on all test points at several timing stages after this trap state can then be displayed. Several different display formats can be selected by the operator using the keyboard. The trap condition is selected initially using the

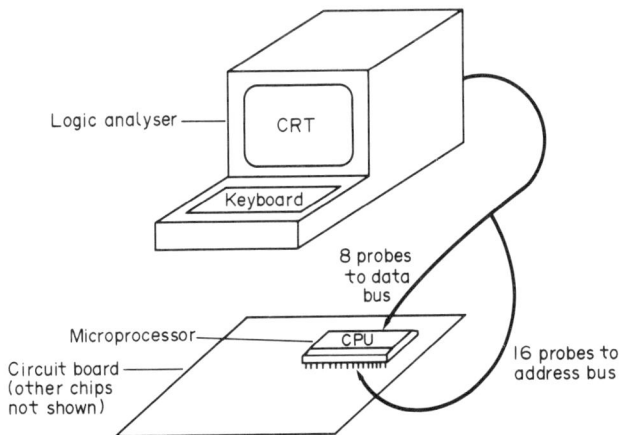

FIG. 14.6. Application of logic analyser.

keyboard, or in earlier models by a series of switches. The analyser can be set to trap once, or to update its memory each time the trap state is encountered.

Figure 14.7 shows several CRT display formats which can be selected with most modern logic analysers. The displays have the following functions:

(a) *Timing analysis*

In this display mode the logic analyser is basically a multi-beam oscilloscope, and several signal waveforms can be displayed together. This feature can be applied in non-microprocessor systems, and can be used to highlight a fault in a conventional digital gating circuit. Different timebase settings can be selected, as with a CRO.

(b) *State analysis* (binary mode)

In this case the settings of the 16 probes that are connected to the address bus are displayed in the A column, whilst the 8 data bus probe signals are displayed in the B column. The trap state in this case has been set to the settings on the address bus probes displayed on the top line of the display (0110 1011 0010 0000), and each display line indicates the setting of all signals at each timing stage after the trap state was encountered. The operator can page through the analyser's memory to display different settings on the address and data buses following the trap state. In this way program operation can be checked. For example, the top three lines on the display may be the three bus transfers that occurred during the fetch/execute cycle for the three-byte machine code instruction which started at memory address 0110 1011 0010 0000.

(c) *State analysis* (hexadecimal mode)

This display is simply a variation of (b), and it presents the bus settings in hexadecimal form in place of binary. This may be a more useful display if the operator is checking the operation of a machine code program which he has listed in hexadecimal form. The first line of the display indicates the opcode fetch part of the fetch/execute cycle for the instruction at location 6B20—the opcode is 05.

(d) *Map mode*

In this mode the CRT screen represents the full 64 K memory map of the microprocessor. The brightness of illumination of the electron scanning beam at different positions on the CRT represents the frequency with which those memory addresses are encountered. Therefore in the example shown a main program commencing at memory location 6000 loops repeatedly. Additionally smaller programs, which could be subroutines, at 2000 and A000, are entered at different exit points in the main program. The operator

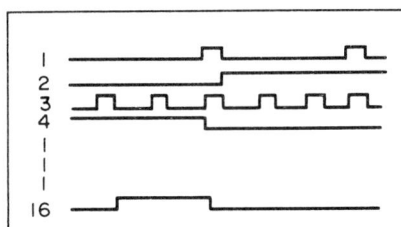

(a) Timing analysis

LINE	A			B	
1	0110	1011	0010 0000	0000	0101
2	0110	1011	0010 0001	0111	0000
3	0110	1011	0010 0010	1001	0010
4	0110	1011	0010 0011	0000	0000
⋮					
16	0010	0000	0000 010	1111	0011

(b) State analysis (binary selected)

LINE	A	B
1	6B20	05
2	6B21	70
3	6B22	92
4	6B23	00
⋮		
16	2005	F3

(c) State analysis (hexadecimal selected)

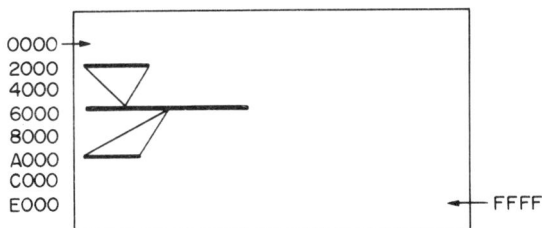

(d) Map mode

FIG. 14.7. Typical logic analyser CRT display formats.

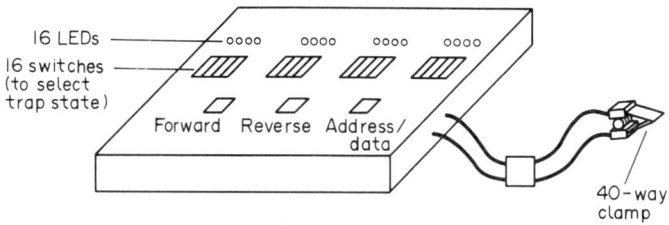

Fig. 14.8. Microbus analyser.

can use this display format to confirm that the correct program sections are executed. This can be useful when fault-finding, e.g. to determine if a section of program which outputs alarm messages is activated, and also during program development.

A simpler and less expensive variation of the logic analyser is the microbus analyser, which is designed to be portable. An example device is illustrated in Fig. 14.8, and it offers only LED indicator display in place of a CRT. The clamp fits over the CPU under test and the operator selects a trap, or "trigger", address in the same manner as for a logic analyser. After the trap state is encountered the operator can step backwards and forwards through the device's memory and display different settings of the address and data buses on the 16 indicating LEDs. This device is suitable for only one type of microprocessor because the pin functions for different microprocessors are not compatible of course.

14.5 SIGNATURE ANALYSER

A signature analyser is an item of test equipment that is only applied with microprocessor systems. A typical signature analyser is illustrated in Fig. 14.9. The device displays a 4-character "signature", which represents signal

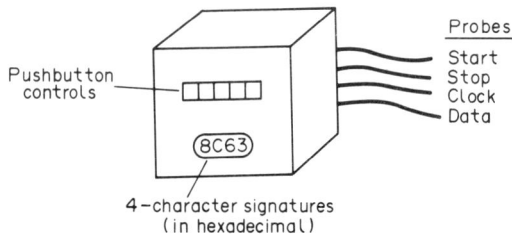

Fig. 14.9. Signature analyser.

activity when the probes are connected to suitable test points in a circuit. The Data probe can be moved around the circuit to measure different signatures.
  The test procedure is as follows:

(a) A test program is activated within the microprocessor circuit under test. This program is identical to that used by the manufacturer when the correct list of signatures was generated. Typically this program repeatedly reads from all memory locations.
(b) The Start, Stop and Clock probes are connected to specified circuit nodes, e.g. Start and Stop may both be connected to an address bus line (perhaps A15) and Clock may be connected to a control bus line (perhaps R/$\overline{W}$).
(c) The Data probe is attached to circuit nodes in the locality of a suspected fault. The signatures at these nodes are compared with a reference list of correct signatures. An incorrect signature can be traced through the circuit until a correct signature is observed. The point in the circuit at which this discontinuity occurs may then indicate a faulty IC or another cause of the fault, e.g. open-circuit.

14.6 CASE STUDIES

14.6.1 Power Supply Failure

Problem. A TTL circuit board, which possesses its own dc power supply, performs the function of a security lock system, i.e. the operator must press a series of pushbuttons in the correct sequence to gain access to a locked room. The system does not respond to the correct operator entry sequence, and tests with a DVM and/or CRO indicate that the +5 V supply to the circuitry is incorrectly set to +8 V.

Cause. The dc regulator IC has failed. It is excessively hot to the touch and an internal short-circuit has occurred—see Fig. 14.10. The IC is replaced and the fault is cleared.

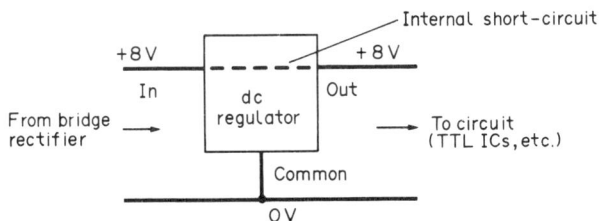

FIG. 14.10. Power supply fault.

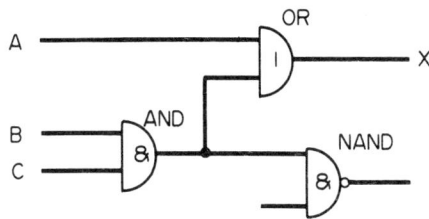

(a) Circuit

A B C	Correct X	Observed X	
0 0 0	0	0	
0 0 1	0	0	
0 1 0	0	0	
0 1 1	1	0 ←	Fault
1 0 0	1	1	
1 0 1	1	1	
1 1 0	1	1	
1 1 1	1	1	

(b) Truth table

Fig. 14.11. Combinational logic system fault.

14.6.2 Logic Gating System Fault

Problem. The combinational logic gating system shown in Fig. 14.11 fails to operate correctly. The operation is tested with a DVM, CRO or logic probe, and the results shown in the truth table apply.

Cause. The only malfunction that occurs is that the output X is set at logic 0 instead of logic 1 when A, B and C are set to 0, 1 and 1. The circuit is tested in this condition, and it is found that the output of the AND gate is 0 in place of 1. This could be caused by one of the following:

(a) the bottom input to the OR gate is fixed at 0,
(b) the input to the NAND gate is fixed at 0,
(c) the output of the AND gate is fixed at 0,
(d) the interconnection from the AND gate to the OR gate is fixed at 0 (short-circuit to ground).

Firstly the OR gate is removed from circuit, and the fault remains—(a) is not the cause. Secondly the NAND gate is removed, and the fault remains—(b) is not the cause. Thirdly the AND gate is removed, and the impedance of the interconnecting track between the AND and OR gates is measured and found to be open-circuit—(d) is not the cause. The AND gate is replaced and the fault disappears (the output of the faulty AND gate was fixed at 0).

14.6.3 Analogue Amplifier Fault

Problem. The non-inverting op-amp circuit of Fig. 14.12 produces a gain of only 1—its design gain is 10.
Cause. The gain of this op-amp stage is:

$$\frac{V_{out}}{V_{in}} = 1 + \frac{R_f}{R_1}$$

$$= 1 + \frac{18K}{2K} = 10$$

It is noticed from this formula that unity gain is achieved theoretically if R_f is zero. This resistor is removed from the circuit and its resistance measured with a DVM. The resistance is only 25 Ω, i.e. a virtual short-circuit. The resistor is replaced and the fault is cleared.

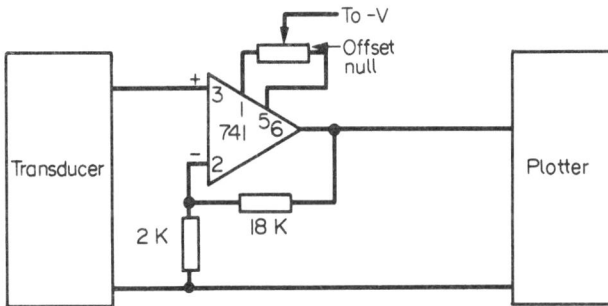

Fig. 14.12. Analogue amplifier.

14.6.4 Data Link Failure

Problem. A microcomputer drives its operator-VDU over a distance of 200 m at 9600 baud using an RS232-C data link (see section 12.3). Approximately half of the characters displayed are corrupted randomly.
Cause. Firstly the VDU is changed, and the fault persists. Secondly the VDU is connected locally to the microcomputer, and the fault disappears. Clearly the problem arises in the transmission cable.

(a) Test connection

(b) Waveforms

(c) Equivalent circuit of interconnecting cable

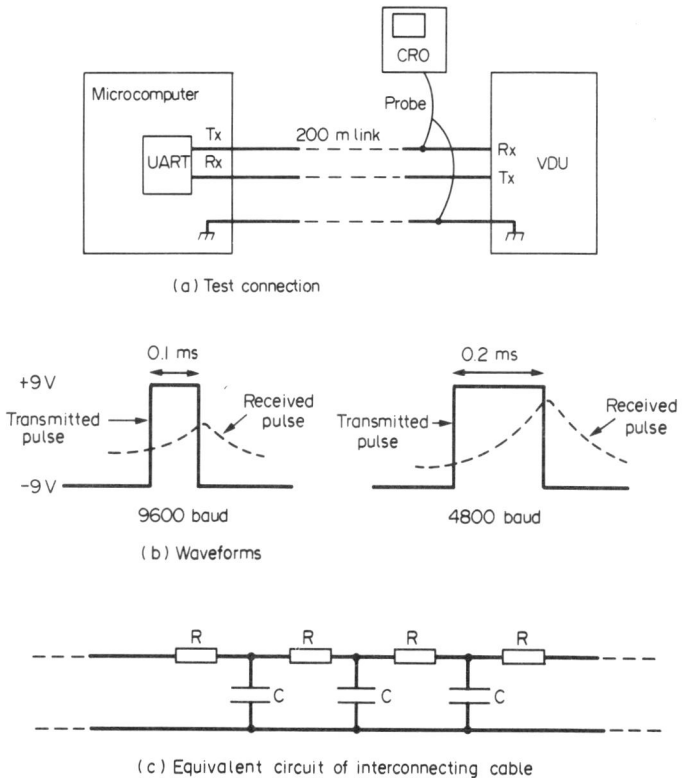

FIG. 14.13. Data link.

A test program that repeatedly outputs the same character is entered into the microcomputer, and the signal at the VDU-end of the transmission cable is measured with a CRO, as shown in Fig. 14.13. The transmitted pulse is severely rounded by the RC (Resistance Capacitance) effect of the interconnecting cable, as shown in diagram (b). The received pulse is of insufficient magnitude to trigger the VDU input circuit reliably, and bits are effectively being "dropped" in the transmission line. The problem is cleared by halving the baud rate at both ends of the data link. Alternative measures that may clear this problem are:

(a) replace cable with one with a better specification, e.g. lower capacitance,
(b) use RS422 (current drive) in place of RS232-C,
(c) employ Schmitt trigger circuits at each end of the link.

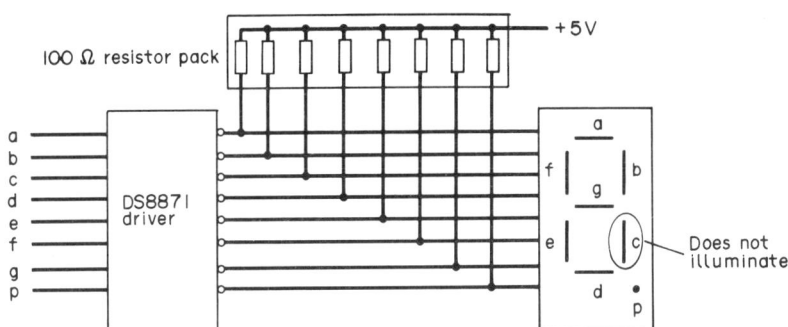

Fig. 14.14. Segment display system

14.6.5 Segment Display Fault

Problem. One segment (c) in a 7-segment display does not light.
Cause. The circuit arrangement is shown in Fig. 14.14. The cause could
be failure of:

(a) segment LED itself,
(b) 100 Ω resistor in resistor-pack,
(c) driver.

All segment bits are set to logic 1 at the input to the driver in order to test
the circuit, and application of a logic probe or DVM or CRO to the driver
outputs indicates that all bits are 0 except segment c, which is at logic 1. The
driver IC is replaced and the fault is cleared.

14.6.6 Microcomputer Data Logger Fault

Problem. A microcomputer, which performs a data logging task, should
light an alarm indicator when a thermostat, which measures motor bearing
temperature, is set. The indicator does not light when the bearing temperature is known to be high.
Cause. A small test program is inserted into the microcomputer to light
the alarm indicator. The indicator does light, and so correct operation of the
output circuit is verified. The microcomputer input circuit for the thermostat, or the data logging program itself, is now suspected of incorrect
operation.
Probes from a logic analyser are now connected to suitable access points

FIG. 14.15. Logic analyser display for microcomputer data logger fault.

on the microprocessor's address and data buses. A trap state is selected for the logic analyser at the memory address of the program instruction that reads in the setting of the thermostat through an input port. Observation of the displayed setting of the data bus, as shown in Fig. 14.15, indicates that all 0s are read in, even when the bearing temperature is high—a 1 should be read in. A wire short-circuit is placed across the input terminals from the thermostat to simulate the setting of the signal, and the test is repeated. This time a 1 is read in. The fault is that the thermostat is not being set correctly. It is replaced and the fault is cleared.

14.6.7 Program Failures in Office Computer

Problem. Small programs in a disk-based microcomputer system run successfully, but large programs "crash" (fail to run correctly).

Cause. It is suspected that a problem exists in the upper parts of the microcomputer's main memory, which is used by large, but not by small,

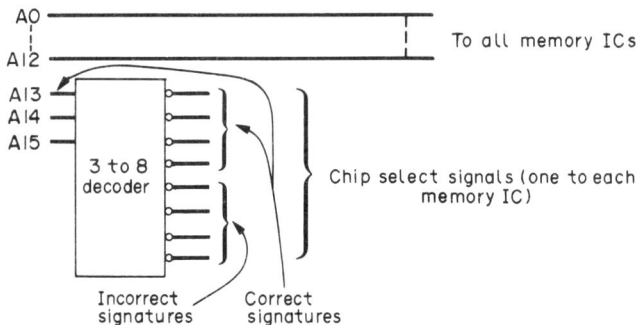

FIG. 14.16. Memory address decoding for disk-based microcomputer.

programs. A signature list exists for this microcomputer configuration, and a signature analyser is available. The appropriate test program is run, and signatures are checked in the vicinity of the address decoding circuit and on memory chips associated with the upper half of the memory address range, as shown in Fig. 14.16. Incorrect signatures are detected on the bottom four outputs from the 3 to 8 decoder IC, whilst signatures on the three inputs are correct. The decoder chip is replaced, and the fault is cleared.

BIBLIOGRAPHY

1. *Troubleshooting on Microprocessor Systems*, G. B. Williams, Pergamon, 1984.
2. *Digital Electronics*, Robert C. Genn, Jr., Prentice/Hall, 1982.

EXERCISES

1. MTBF is defined in the text. What do you think the abbreviation MTTR means?
2. What is a "dry joint"?
3. List different types of circuit faults that can be located by simple observation.
4. What is the facility of "pulse stretching" on a logic probe?
5. List the three types of signals that can be generated by a logic pulser.
6. Which item of test equipment indicates the memory area in which the main program is active?
7. Why is a microbus analyser normally only suitable for a single type of microprocessor?
8. Which items of test equipment require that the user should possess some knowledge of machine code programming?
9. What procedure would you use with a logic probe and logic pulser to locate a fault in a section of memory circuit that employs a 2 to 4 decoder and RAM chip?
10. Write out the CRT display of a logic analyser set to hexadecimal display mode for the following Z80 program, which commences at memory address hex. 0200:

```
LD    B,3
LD    A,0FH
OUT   (50H),A
JP    0300H
```

11. A microcomputer repair workshop purchases a signature analyser to assist fault-finding with a particular home computer. What procedures are required to generate a list of correct signatures for later use with faulty machines?

12. List typical fault symptoms that occur with faulty dc power supplies.

13. What are likely causes of the following faults? Suggest items of test equipment and a test procedure that you would apply to help to locate each fault.

   (a) there is no activity on a microprocessor's address and data buses.
   (b) all items of equipment fed from a microcomputer PIO are inoperative.
   (c) an analogue instrument signal that is displayed by a microcomputer is faulty (permanently zero).
   (d) a test program indicates that every alternative block of 8 memory locations throughout a 32 k RAM memory board cannot be accessed successfully.
   (e) a microcomputer board that is applied to program EPROMs operates correctly (displays bytes to be programmed successfully) but does not actually program bytes.
   (f) a microcomputer system that should print an alarm message on a dedicated printer whenever a plant contact-closure (connected as an interrupt) occurs fails to print.

Appendix A—TTL Range of ICs

(Reprinted by courtesy of RS Components Limited)

Issued November 1984 **5667**

RS data — 74 Series logic families

This data sheet gives the pin connections and availability for the five 74 series logic families offered.

Standard 74 Series TTL

A range of popular transistor—transistor logic integrated circuits for use in basic circuits. 5 V dc supply, 0°C to +70°C operating temperature range.

Low power Schottky 74 LS Series TTL

A Schottky process using shallower diffusions yields devices with a five fold decrease in power consumption and an increase in speed compared to standard 74 TTL. 5 V dc supply, 0°C to +70°C operating temperature range.

Advance low power Schottky 74 ALS TTL

Advanced low power Schottky TTL devices are directly compatible with LS and standard TTL devices. An advanced oxide isolated construction uses small geometries giving approximately twice the speed with half the power consumption of 74 LS TTL. 5 V dc supply, 0°C to +70°C operating temperature range.

High speed CMOS 74 HC and 74 HCT

High speed CMOS devices are fabricated using the latest technology oxide isolated CMOS process giving good speed performance comparable with LS TTL but with much lower power consumption of the same order as 4000 CMOS.

Two families are offered 74 HC with CMOS compatible inputs for high noise immunity and 74 HCT with directly TTL compatible inputs. Both families outputs are compatible with TTL and CMOS. 74 HC devices work on a 2 to 6 V dc supply with 74 HCT types functioning with a 4.5 to 5.5 V dc supply. Operating temperature range for both families is a wide −40°C to +85°C.

Connections shown are top view. A "negation" circle at any output or input within the schematic indicates that the terminal is active LOW or at clocking inputs the device is negative edge triggered.

Unused inputs should be connected to the appropriate defined logic level in order to achieve output conditions in line with the device truth table. As with standard CMOS high speed CMOS unused inputs must be connected to defined logic levels. For active HIGH inputs with standard TTL and ALS TTL a pull up resistor to V_{CC} should be used; up to 25 unused inputs can be connected to each resistor. With LS TTL, HC and HCT unused active HIGH inputs can be directly tied to +ve supply so long as the connecting leads are short and the supply is adequately decoupled. Unused active LOW inputs can be directly connected to ground with all families.

Abbreviations used throughout this data sheet

A, B, C, D and E	Data inputs binary weight (where applicable)
	A = 1; B = 2; C = 4; D = 8; E = 16
a, b, c, d, etc	Segment outputs on 7-segment decoder drive
BCD	Binary coded decimal
BI	Blanking input
$C_{in, out}$	Carry in or out
CEP	Count enable parallel input

CER	Count enable ripple input
CK	Clock
CS	Chip select
D, JK	Data input to flip-flops
EN	Enable
GND	Ground 0 V terminal
I/O	Input/Output
LT	Lamp test
MR	Master test
OEN	Output enable
PE	Parallel enable (active low) input
Q	Output, may have a letter indicating weighting
RBI	Ripple blanking input
RBO	Ripple blanking output
RC, C, R	Capacitor and resistor timing on monostables
RCO	Ripple carry output
S	Sum output
SDL	Serial data in left shift
SDR	Serial data in right shift
SQ	Serial output
SR	Synchronous reset
TC	Terminal count output
V_{CC}	+ supply terminal
$\sqcap$	Schmitt device or function

WARNING!

ESD SENSITIVE DEVICE

CAUTION: HC and HCT devices

ESD (Electro-Static-Discharge) sensitive device. The digital inputs are diode protected; however, permanent damage may occur on unconnected devices subject to high energy electrostatic fields. Unused devices must be stored in conductive foam or shunts. The protective foam should be discharged to the destination socket before devices are removed.

HC, HCT, LS, TTL and 4000 CMOS characteristics

	Metal gate CMOS	LS TTL	HCT	HC
Supply voltage range (V$_{CC}$)	3-15V	5V±5%	5V±10%	2-6V
Typical quiescent dissipation per gate	2.5 nW	2mW	2.5 nW	1μW
Max. quiescent current per package at 85°C	7.5 μA	3 mA	20 μA	20 μA
Typical power dissipation per gate (V$_{CC}$=5V, C$_L$ = 50 pF) at 10 kHz at 100 kHz at 1 MHz at 10 MHz	25 μW 250 μW 2.5 mW –	2 mW 2 mW 2.8 mW 12.5 mW	14 μW 140 μW 1.4 mW 14 mW	14 μW 140 μW 1.4 mW 14 mW
Fan-out (TTL loads)	1	10	10	10
Typical propagation delay C$_L$= 15 pF C$_L$= 100 pF	90ns 175 ns	10ns 17 ns	9ns 11.5 ns	8ns –
Max. operating frequency width C$_L$=15 pF	3MHz	25MHz	25MHz	40MHz
Operating temperature range °C	–40 to +85	0 to+70	–40 to +85	–40 to +85

High speed CMOS comparisons

HC logic implements TTL equivalent functions with the same pin outs as TTL. HC is not designed to be directly plug-in replaceable, but, with care, some TTL systems can be converted to 74HC with little or no modification. The replaceability of HC is determined by several factors.

One factor is the difference is input levels. In systems where all TTL is not being replaced and TTL outputs feed CMOS inputs, the input high voltages, as specified, are not totally compatible. Although TTL outputs will typically drive HC inputs correctly, an external pull-up resistor should be added to the TTL outputs, or a 74HCT TTL compatible circuit should be used.

HCT inputs are directly compatible with TTL outputs, and HCT outputs are completely compatible with the various TTL family's input specifications; therefore, there is no problem when HCT and TTL is mixed. Another source of possible problems can occur when the LS design floats device inputs. This practice is not recommended when using LS-TTL, but it is sometimes done. Usually, TTL inputs float high; however, CMOS inputs may float either high or low depending on the static charge on the input. It is therefore important to always tie unused CMOS inputs to either V$_{CC}$ or ground to avoid incorrect logic functioning.

A third factor to consider when replacing any TTL logic is ac performance. The logic functions provided by 74 HCT are equivalent to LS-TTL, and the propagation delay, set-up and hold times are similar to LS. However, there are some differences in the way CMOS circuits are implemented which will cause differences in speed. For the most part, these differences are minor, but it is important to verify that they do not affect the design.

Conclusion

The 74 HC families represent a major step forward in CMOS performance. They are capable of being designed into virtually any application which now uses LS-TTL with substantial improvement in power consumption, or used exclusively for high performance new designs. ALS offers faster speeds than HCMOS, but still does not have the input and output advantages or the lower power consumption of CMOS. Because of its high input impedance and large output drive, HC logic is actually easier to use.

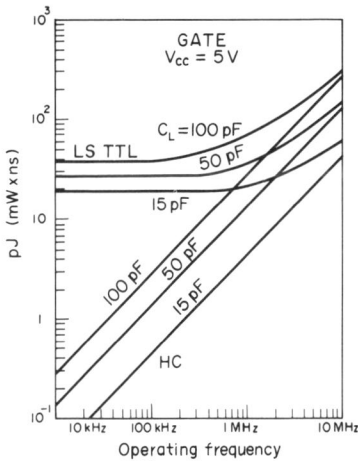

FIG. A1. Power-speed product for HCMOS and LSTTL gate shows the advantage of HCMOS at frequencies below 10 MHz.

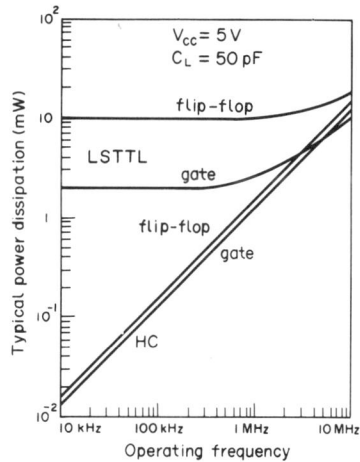

FIG. A2. The power cross-over frequency is about 5 MHz for a single gate and above 10 MHz for a single flip-flop.

10^3

10 FLIP-FLOPS V_{CC} = 5V
I loaded with 50 pF
9 loaded with 15 pF

10^2

LSTTL

Power dissipation (mW)

10

1

HC

10^{-1}

100 kHz   1 MHz   10 MHz   100 MHz

operating frequency

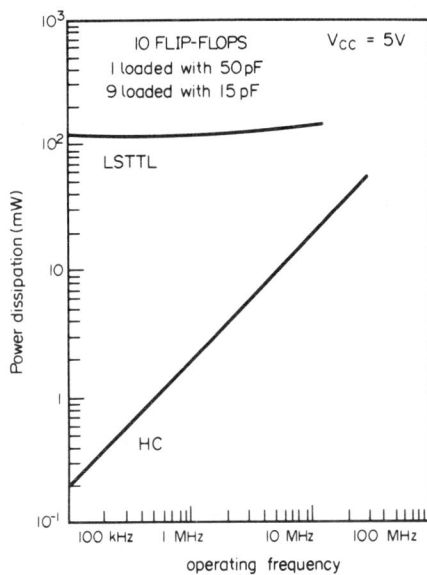

Fig. A3. A simulated MSI circuit shows distinct power saving at all frequencies.

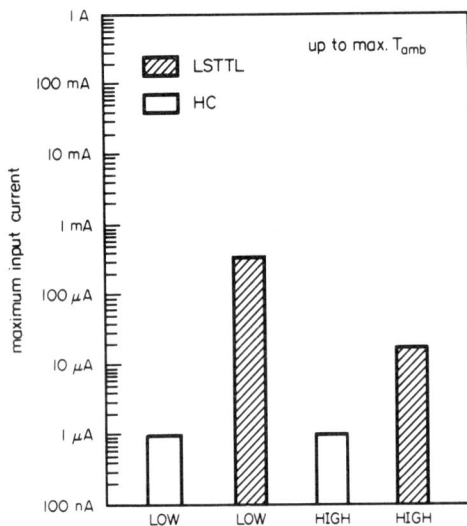

1 A

up to max. T_{amb}

100 mA   LSTTL

HC

10 mA

maximum input current

1 mA

100 μA

10 μA

1 μA

100 nA   LOW   LOW   HIGH   HIGH

Fig. A4. Input current for HCMOS is symmetrical and much lower than that of LSTTL. Theoretically, one HCMOS output can drive nearly a thousand inputs, but capacitance considerations will probably predominate.

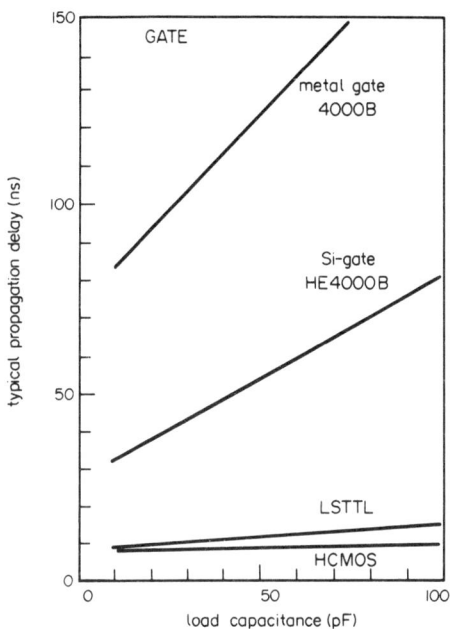

FIG. A5. As its name suggests high speed CMOS is fast. Gate propagation delay with 50 pF load is $\frac{1}{6}$th that of earlier silicon gate CMOS and $\frac{1}{12}$th that of metal gate CMOS.

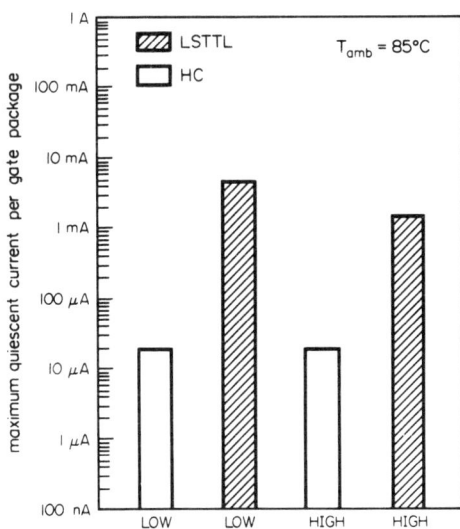

FIG. A6. Even a total HCMOS package of four gates under worst case static conditions consumes more than two orders of magnitude less power than an equivalent LSTTL gate package.

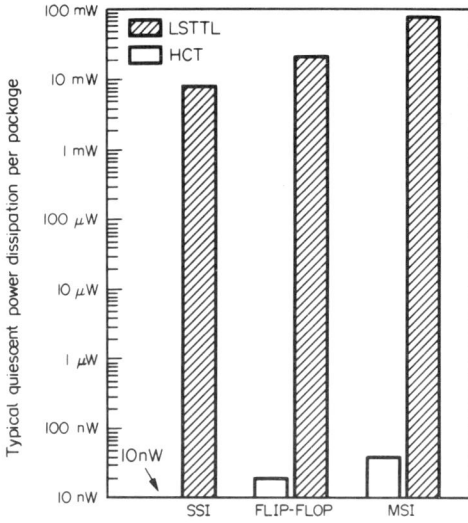

FIG. A7. Unlike LSTTL circuits, HCMOS circuits consume very little power when they are not switching.

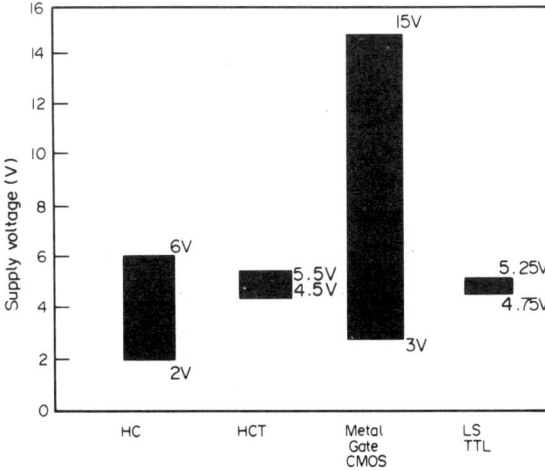

FIG. A8. Comparison of supply voltage range.

To summarise, the outstanding features of high-speed CMOS are:

● Low power dissipation. Typical quiescent current per package is 2 nA for gates, 4 nA for flip-flops and 8 nA for MSI. For gates, maximum quiescent current per package at 85°C is 20 μA. Typical gate operating current is 3 μA at 10 kHz, 30 μA at 100 kHz and 300 μA at 1 MHz with a 5 V supply. This compares with LS TTL operating currents of 400 μA up to 100 kHz and 560 μA at 1 MHz for each gate.

● Typical operating frequency up to 50 MHz (15 pF, 25°C). With a 5 V supply, typical propagation delay for a gate is 9 to 11.5 ns for either HIGH to LOW or LOW to HIGH transitions into capacitive loads of between 15 and 100 pF, This is a fifth of the gate propagation delay for earlier silicon-gate CMOS.

● Functions and pinning identical to popular LS TTL.

● TTL input switching level HCT devices. These circuits operate from a 5 V ±10% supply and are mainly for use as pin-compatible CMOS replacements for LS TTL to reduce power consumption without loss of speed. These types are also suitable for converting switching levels from TTL to CMOS. CMOS input switching level HC devices, with an operating supply voltage of 2 to 6 V, are ideal for wide noise margin new designs.

● Fan-out of 10 LS TTL loads (4 mA) for standard outputs, 15 LS TTL loads (6 mA) for bus driver outputs. This is ten times more drive capability than earlier CMOS circuits.

● Standardised output buffers allow symmetrical output current sourcing and sinking for equal output rise and fall times (7.5 ns for standard outputs and 6 ns for bus driver outputs). This results in simplified design combined with optimum speed and performance.

● High immunity to electrostatic discharges.

● Wide operating temperature range: −40°C to +85°C.

● Virtually latch-up free.

00 Quadruple 2-input NAND gate

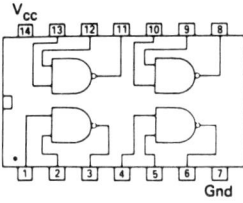

V_{cc}

Gnd

01 Quadruple 2-input NAND gate with open collector output

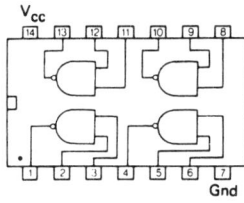

V_{cc}

Gnd

02 Quadruple 2-input NOR gate

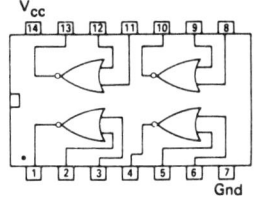

V_{cc}

Gnd

03 Quadruple 2-input NAND gate — open collector inputs

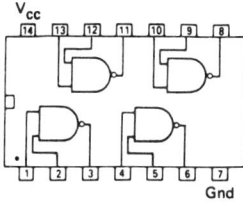

V_{cc}

Gnd

04 Hex inverter

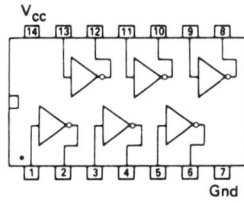

V_{cc}

Gnd

05 Hex inverter-open collector outputs

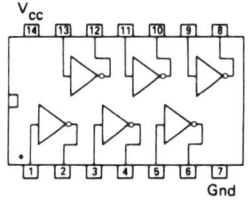

V_{cc}

Gnd

06 Hex inverter with high voltage open collector output

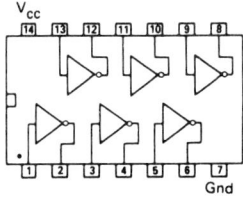

V_{cc}

Gnd

07 Hex driver with open collector output

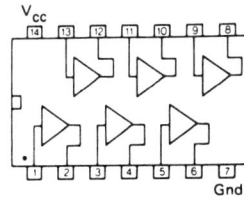

V_{cc}

Gnd

08 Quadruple 2-input AND gate

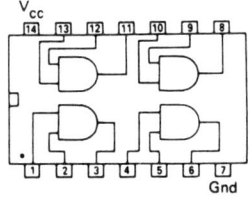

V_{cc}

Gnd

09 Quad 2-input AND gate-open collector outputs

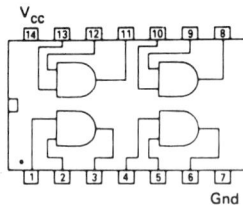

V_{cc}

Gnd

10 Triple 3-input NAND gate

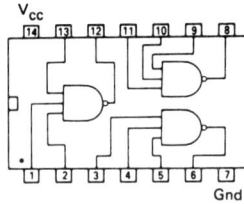

V_{cc}

Gnd

11 Triple 3-input AND gate

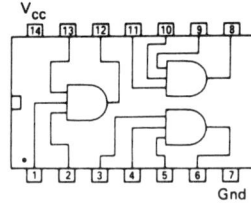

V_{cc}

Gnd

13 Dual 4-input NAND gate Schmitt trigger

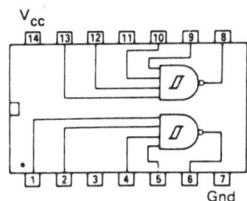

14 Hex Schmitt Trigger

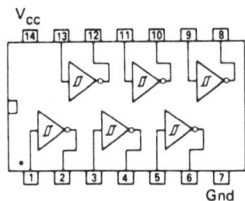

15 Triple 3-input AND gate — open collector outputs

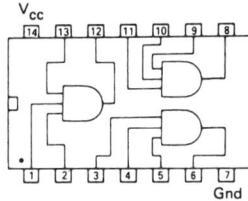

16 Hex inverter with open collector output

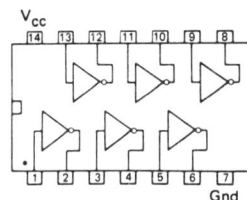

20 Dual 4-input NAND gate

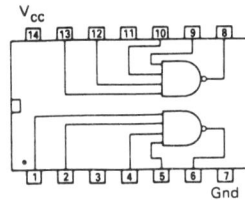

21 Dual 4-input AND gate

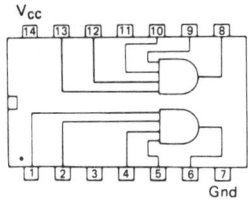

22 Dual 4-input NAND gate — open collector outputs

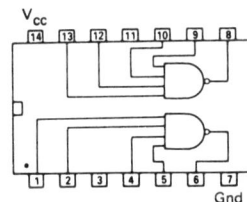

25 Dual 4-input NOR gate with strobe

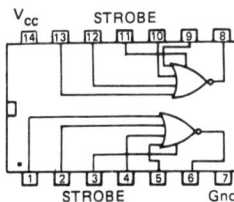

26 Quad 2-input NAND buffer-open collector outputs

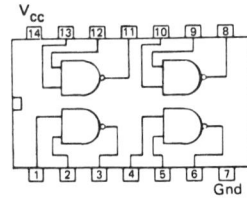

27 Triple 3-input NOR gate

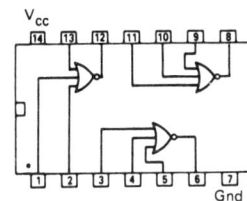

28 Quad 2-input NOR buffer

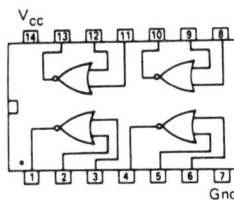

30 8-input NAND gate

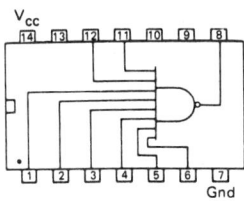

32 Quadruple 2-input OR gate

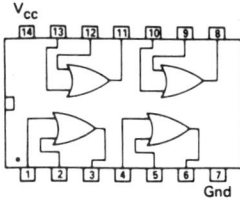

33 Quad 2-input NOR buffer-open collector outputs

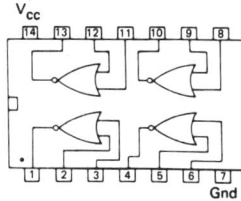

37 Quadruple 2-input NAND buffer

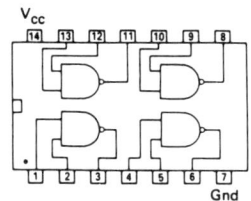

38 Quadruple 2-input NAND buffer — open collector outputs

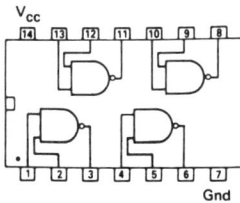

40 Dual 4-input NAND buffer

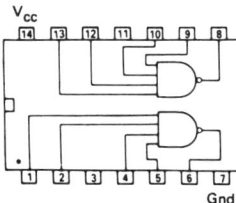

42 BCD-to-decimal decoder

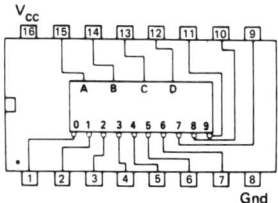

45 BCD-to-decimal decoder/driver

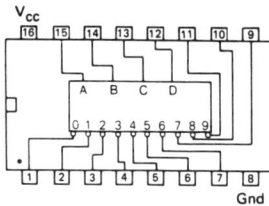

47 BCD-to-7 segment decoder/driver — open collector outputs

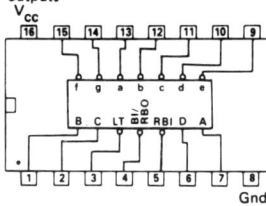

48 BCD-to-7 segment decoder/driver

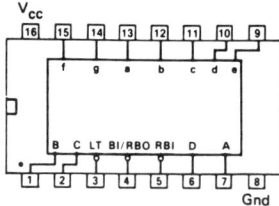

49 BCD-to-7-segment decoder/driver — open collector outputs

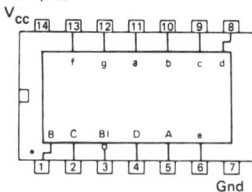

51 Dual 2-wide 2-input/3-input AND-OR-INVERT gate

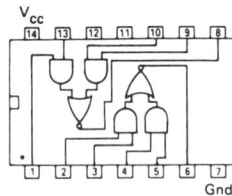

54 3-2-2-3 input AND-OR-INVERT gate

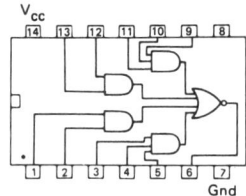

55 2-wide 4-input AND-OR-INVERT gate

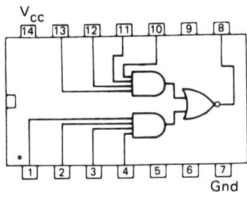

70 J-K flip-flop

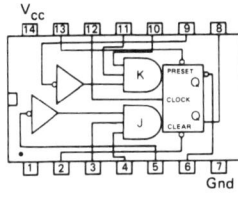

72 J-K master-slave flip-flop

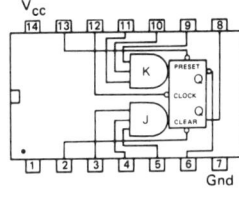

73 Dual JK negative edge-triggered Flip-Flop

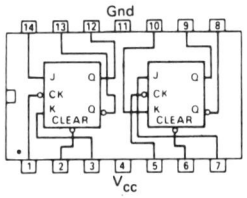

74 Dual D-type edge-triggered Flip-Flop

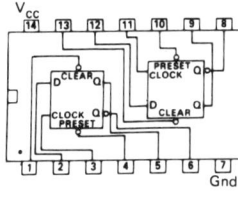

75 4-bit D Latch

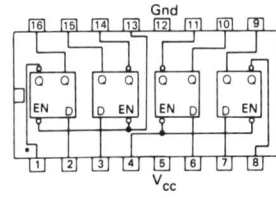

76 Dual JK Flip Flop with set and clear

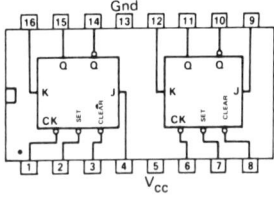

78 Dual JK Flip-Flop

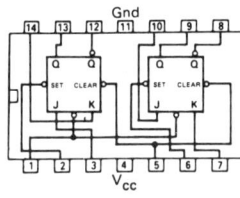

83A 4-bit Binary full adder

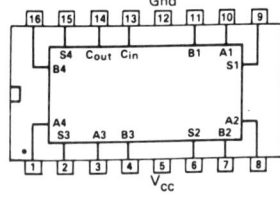

85 4-bit magnitude comparator

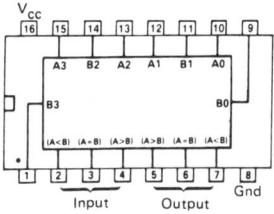

86 Quadruple 2-input exclusive OR gate

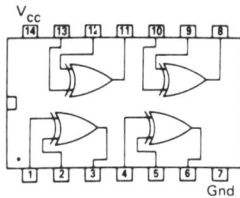

90 Decade counter

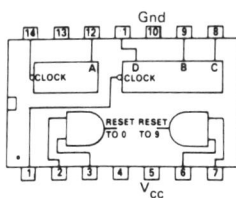

92 Divide-by-twelve counter

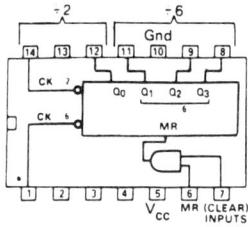

93 4-bit binary counter

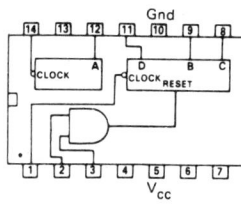

95B 4-bit shift register

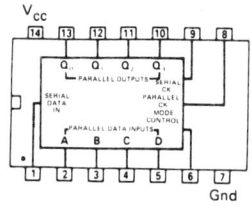

96 5-bit shift register

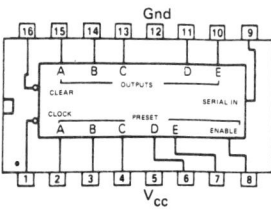

107 Dual JK Flip-Flop

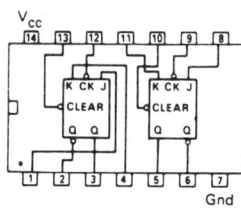

109 Dual JK positive edge-triggered Flip-Flop

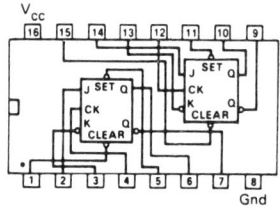

112 Dual JK edge-triggered flip-flop

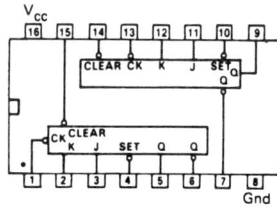

113 Dual JK negative edge-triggered Flip-Flop

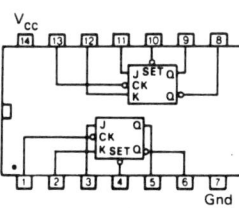

114 Dual JK negative edge-triggered Flip-Flop

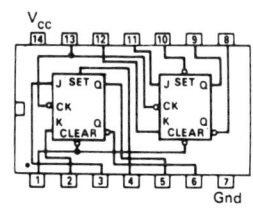

121 Monostable Multivibrator

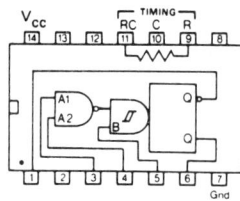

123 Dual monostable - retriggerable

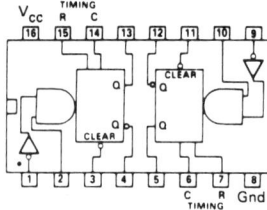

125 Quad 3-state buffer (active low enable)

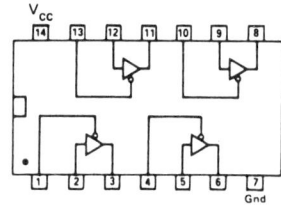

126 Quad 3-state buffer (active high enable)

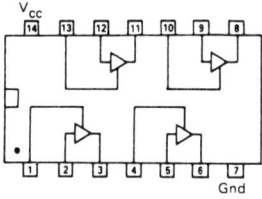

128 Quad line driver

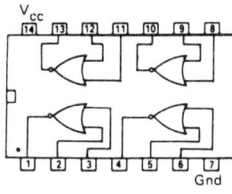

132 Quadruple 2-input NAND Schmitt gate

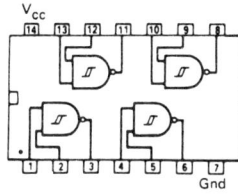

133 13-input NAND gate

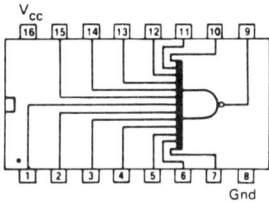

137 3-line to 8-line Decoder/Demultiplexer with address latches

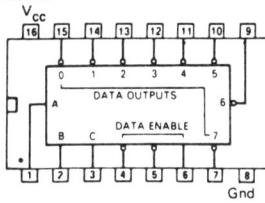

138 3 to 8 line Decoder/Multiplexer

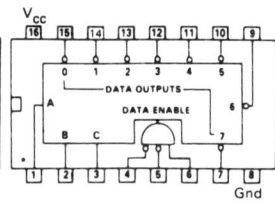

139 Dual 1 of 4 Decoder

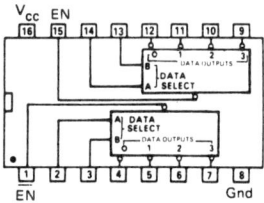

141 BCD-to-decimal decoder driver

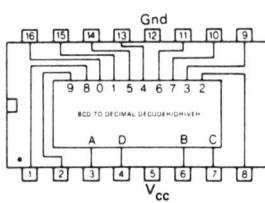

145 BCD-to-decimal decoder/driver

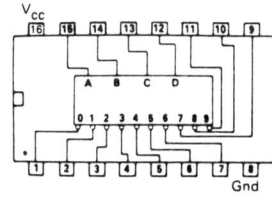

148 Octal priority encoder 8 line to 3 line

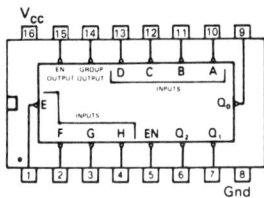

151 1 of 8 Data Selector/Multiplexer

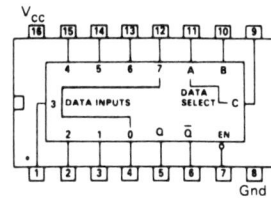

153 Dual 4-line to 1-line Data Selectors/Multiplexers

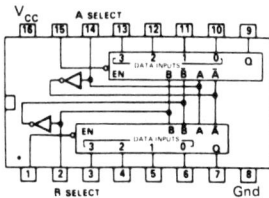

154 4 to 16 line Decoder

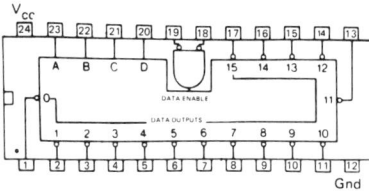

155 Dual 1 of 4 Decoder/Demultiplexer

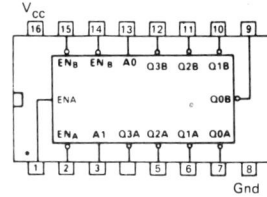

156 Dual 1-of-4 Decoder/Demultiplexer with open collector outputs

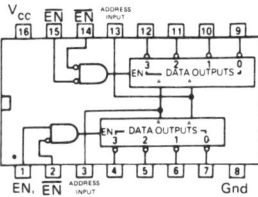

157 Quad 2 to 1-line Data Selectors/Multiplexers

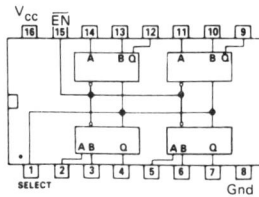

158 Quad 2 to 1-line Data selectors/Multiplexers with inverted outputs

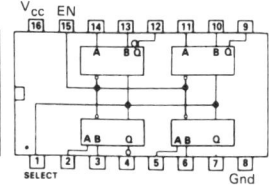

160 BCD decade counter – asynchronous reset

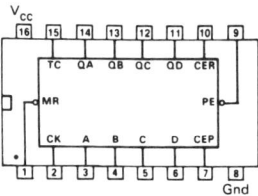

161 Binary counter – asynchronous reset

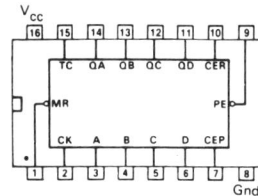

162 BCD counter – synchronous reset

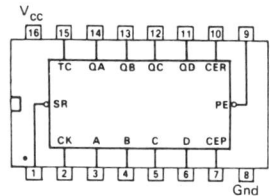

163 Binary counter · synchronous reset

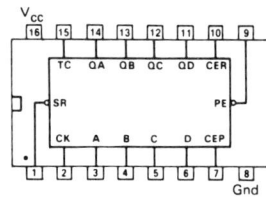

164 Serial-in parallel-out shift register

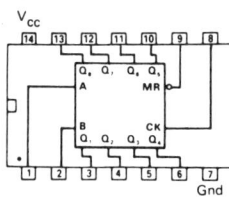

165 8-bit parallel to serial converter

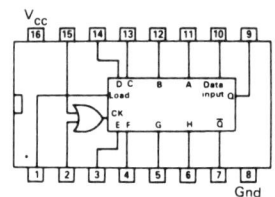

174 Hex D-type Flip-Flops

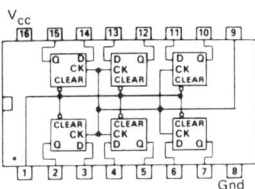

175 Quad D-type Flip-Flops

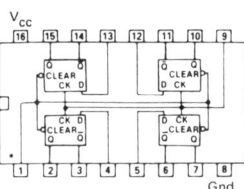

180 Parity generator/checker 9-bit odd/even

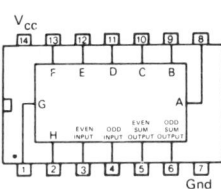

181 4-bit arithmetic logic unit

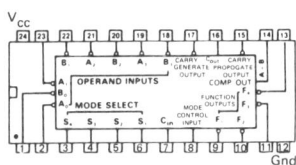

191 Binary synchronous up/down counter

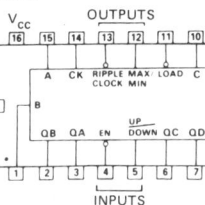

192 Up/Down decade counter — with preset inputs

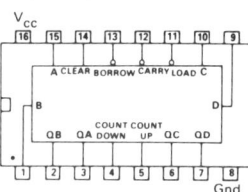

193 Up/Down binary counter-with preset inputs

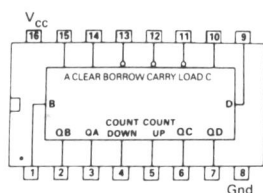

194 A 4-bit bidirectional universal shift register

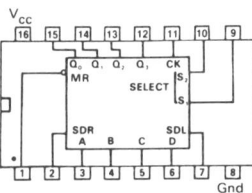

195 4-bit parallel-access shift register

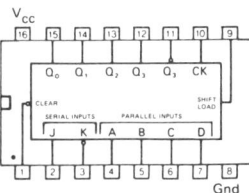

196 4-stage presettable ripple counter

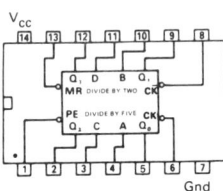

197 Presettable binary ripple counter

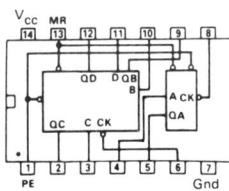

221 Dual monostable multivibrator

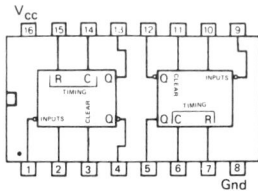

ICM-0

240 Octal buffer - three state inverting

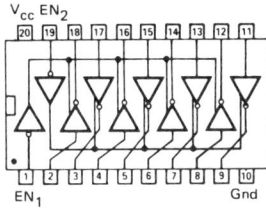

V_{cc} EN_2
EN₁   Gnd

241 Octal buffer — three state non-inverting

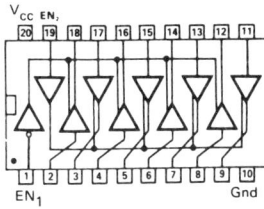

V_{cc} EN₂
EN₁   Gnd

242 Quad bus transceiver — inverting

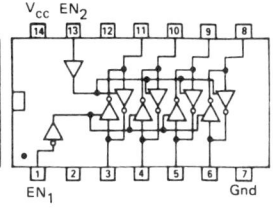

V_{cc} EN_2
EN₁   Gnd

243 Quad bus transceiver — non-inverting

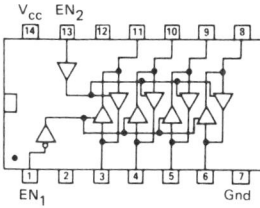

V_{cc} EN_2
EN₁   Gnd

244 Octal buffer — three state non-inverting

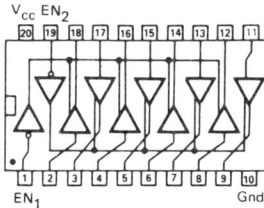

V_{cc} EN_2
EN₁   Gnd

245 Octal bus transceiver with 3 state outputs

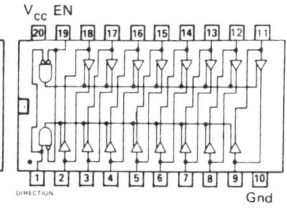

V_{cc} EN
DIRECTION   Gnd

251 1 of 8 Data selector/Multiplexer with 3 state outputs

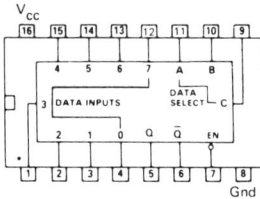

V_{cc}
Gnd

253 Dual 4 - input multiplexer with 3 state outputs

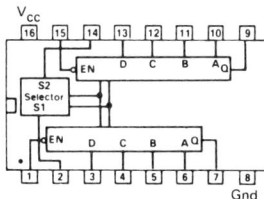

V_{cc}
Gnd

256 Dual 4-bit addressable latch

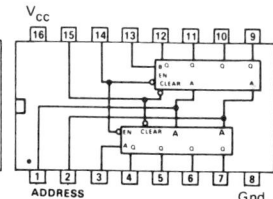

V_{cc}
ADDRESS INPUTS   Gnd

257 Quad 2-input multiplexer with 3 state outputs

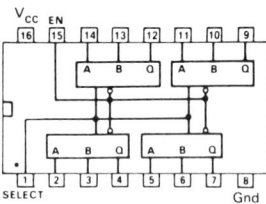

V_{cc} EN
SELECT   Gnd

258 Quad 2-input multiplexer with 3 state outputs

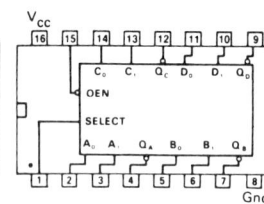

V_{cc}
Gnd

259 8-bit addressable latch

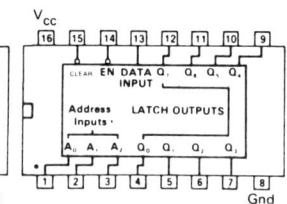

V_{cc}
Gnd

266 Quad 2-input Exclusive NOR gate **273** 8-bit register with clear **283** 4-bit binary full adder

298 Quad 2-port register (Quad 2-input multiplexer with storage) **299** 8-bit universal shift/storage register with common parallel/0 pins 3 state **321** Crystal controlled oscillator

323 8-bit universal shift/storage register with synchronous reset and common 1/0 pins 3 state **352** Dual 4-input multiplexer inverting **353** Dual 4-input multiplexer with 3 state outputs inverting

354 8-line to 1-line data selector/multiplexer/register **356** 8-line to 1-line data selector/multiplexer/register **365** Hex 3 state buffer non-inverting

366 Hex 3 state buffer inverting

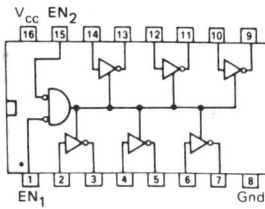

367 Hex 3-state buffer

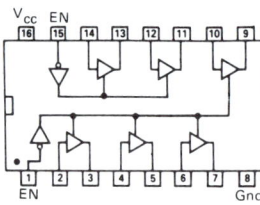

368 Hex 3-state inverter) buffer
(separate 2-bit & 4-bit sections)

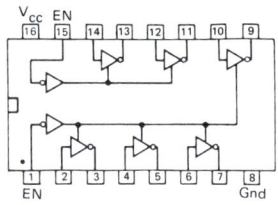

373 Octal transparent latch with 3 state outputs

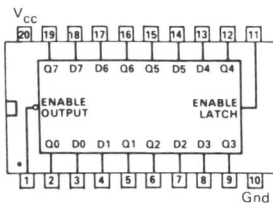

374 Octal D-type flip-flop with 3 state outputs

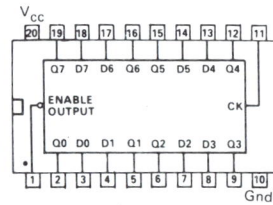

377 Octal D-type flip-flop with enable

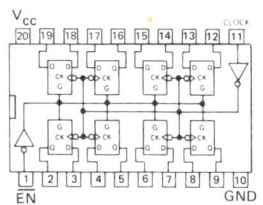

378 Hex D register

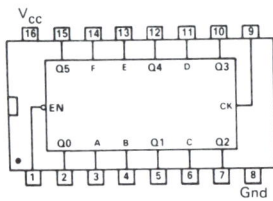

390 Dual decade counter

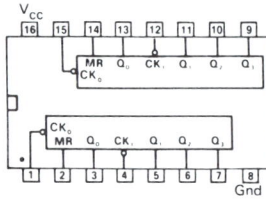

393 Dual 4 stage binary counter

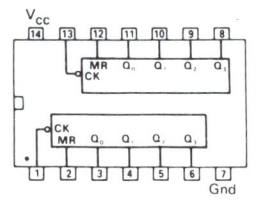

395 4-bit cascadable shift register 3 state

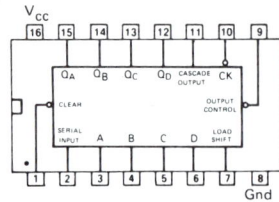

442, 443, 444 Quad tridirectional bus transceivers 3 state

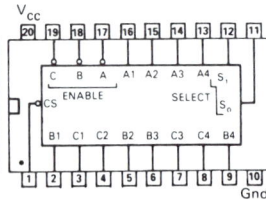

573 Octal D-type transparent latch

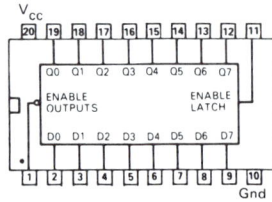

580 Octal D-type transparent latch
inverted outputs

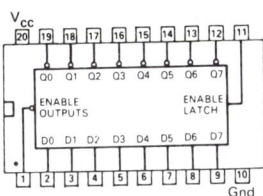

| 20 | 19 | 18 | 17 | 16 | 15 | 14 | 13 | 12 | 11 |

V_{CC}

Q0 Q1 Q2 Q3 Q4 Q5 Q6 Q7
ENABLE OUTPUTS   ENABLE LATCH
D0 D1 D2 D3 D4 D5 D6 D7

| 1 | 2 | 3 | 4 | 5 | 6 | 7 | 8 | 9 | 10 |

Gnd

620 Octal bus transceiver

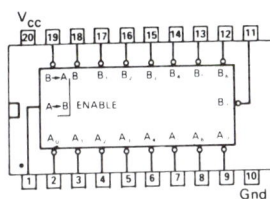

V_{CC}

| 20 | 19 | 18 | 17 | 16 | 15 | 14 | 13 | 12 | 11 |

B→A B_1 B_2 B_3 B_4 B_5
A→B ENABLE   B
A_1 A_2 A_3 A_4 A_5 A_6

| 1 | 2 | 3 | 4 | 5 | 6 | 7 | 8 | 9 | 10 |

Gnd

625 Voltage controlled oscillator

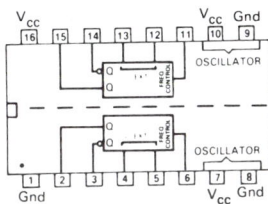

V_{CC} V_{CC} Gnd

| 16 | 15 | 14 | 13 | 12 | 11 | 10 | 9 |

OSCILLATOR

OSCILLATOR

| 1 | 2 | 3 | 4 | 5 | 6 | 7 | 8 |

Gnd V_{CC} Gnd

640 Tri-state, inverting octal
bus transceiver

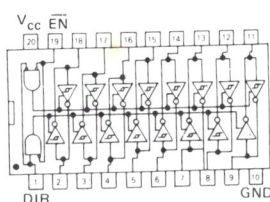

V_{CC} $\overline{EN}$

| 20 | 19 | 18 | 17 | 16 | 15 | 14 | 13 | 12 | 11 |

| 1 | 2 | 3 | 4 | 5 | 6 | 7 | 8 | 9 | 10 |

DIR   GND

643 Tri-state, true and
inverting octal bus
transceiver

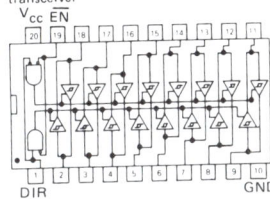

V_{CC} $\overline{EN}$

| 20 | 19 | 18 | 17 | 16 | 15 | 14 | 13 | 12 | 11 |

| 1 | 2 | 3 | 4 | 5 | 6 | 7 | 8 | 9 | 10 |

DIR   GND

669 Up/down binary counter
synchronous

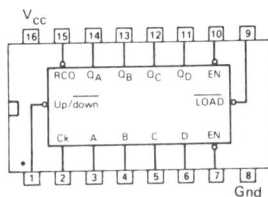

V_{CC}

| 16 | 15 | 14 | 13 | 12 | 11 | 10 | 9 |

RCO Q_A Q_B Q_C Q_D EN
Up/down   LOAD
Ck   A   B   C   D   EN

| 1 | 2 | 3 | 4 | 5 | 6 | 7 | 8 |

Gnd

670 4 x 4 Register file with 3-state
outputs

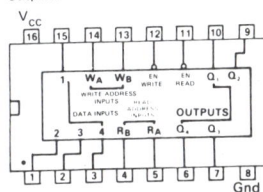

V_{CC}

| 16 | 15 | 14 | 13 | 12 | 11 | 10 | 9 |

W_A W_B EN   EN Q_1 Q_1
WRITE ADDRESS   WRITE READ
INPUTS
DATA INPUTS   OUTPUTS
R_B R_A Q_1 Q_1

| 1 | 2 | 3 | 4 | 5 | 6 | 7 | 8 |

Gnd

673 16-bit shift register, serial to parallel

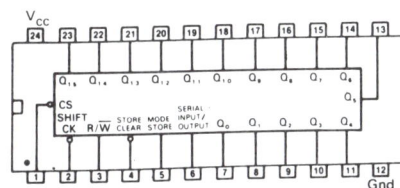

V_{CC}

| 24 | 23 | 22 | 21 | 20 | 19 | 18 | 17 | 16 | 15 | 14 | 13 |

Q_{15} Q_{14} Q_{13} Q_{12} Q_{11} Q_{10} Q_9 Q_8 Q_7 Q_6
                                                                      Q_5
CS
SHIFT   STORE MODE   SERIAL
CK R/W CLEAR STORE OUTPUT Q_0 Q_1 Q_2 Q_3 Q_4

| 1 | 2 | 3 | 4 | 5 | 6 | 7 | 8 | 9 | 10 | 11 | 12 |

Gnd

674 16-bit shift register, parallel to serial

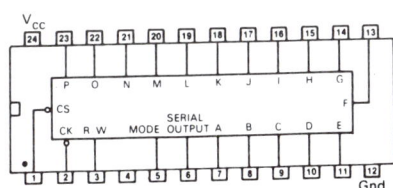

V_{CC}

| 24 | 23 | 22 | 21 | 20 | 19 | 18 | 17 | 16 | 15 | 14 | 13 |

P O N M L K J I H G
                                        F
CS
CK R W   MODE SERIAL A B C D E
           OUTPUT

| 1 | 2 | 3 | 4 | 5 | 6 | 7 | 8 | 9 | 10 | 11 | 12 |

Gnd

682 8-bit magnitude comparator

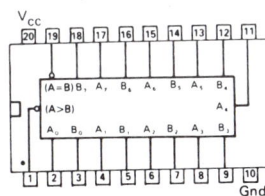

V_{CC}

| 20 | 19 | 18 | 17 | 16 | 15 | 14 | 13 | 12 | 11 |

(A=B)B_5 A_5 B_6 A_6 B_7 A_7 B_8
(A>B) A_8
A_0 B_5 A_1 B_2 A_3 B_1 A_3 B_1

| 1 | 2 | 3 | 4 | 5 | 6 | 7 | 8 | 9 | 10 |

Gnd

Standard LS, ALS, HC and HCT availability guide

No.	STD	LS	ALS	HC	HCT
00					
01					
02					
03					
04					
05					
06					
07					
08					
09					
10					
11					
13					
14					
15					
16					
20					
21					
22					
25					
26					
27					
28					
30					
32					
33					
37					
38					
40					
42					
45					
47					
48					
49					
51					
54					

No.	STD	LS	ALS	HC	HCT
55					
70					
72					
73					
74					
75					
76					
78					
83					
85					
86					
90					
92					
93					
95					
96					
107					
109					
112					
113					
114					
121					
123					
125					
126					
128					
132					
133					
137					
138					
139					
141					
145					
151					
148					
153					

No.	STD	LS	ALS	HC	HCT
154					
155					
156					
157					
158					
160					
161					
162					
163					
164					
165					
174					
175					
180					
181					
191					
192					
193					
194					
195					
196					
197					
221					
240					
241					
242					
243					
244					
245					
251					
253					
256					
257					
258					
259					
266					

No.	STD	LS	ALS	HC	HCT
273					
283					
298					
299					
321					
323					
352					
353					
354					
356					
365					
366					
367					
368					
373					
374					
377					
376					
390					
393					
395					
442					
443					
444					
573					
580					
620					
625					
640					
643					
669					
670					
673					
674					
682					

Appendix B—ASCII Code

Character	Hex.	Character	Hex.	Character	Hex.
NUL	00	0	30		60
SOH	01	1	31	a	61
STX	02	2	32	b	62
ETX	03	3	33	c	63
EOT	04	4	34	d	64
ENQ	05	5	35	e	65
ACK	06	6	36	f	66
BEL	07	7	37	g	67
BS	08	8	38	h	68
HT	09	9	39	i	69
LF	0A	:	3A	j	6A
VT	0B	;	3B	k	6B
FF	0C	<	3C	l	6C
CR	0D		3D	m	6D
S0	0E	>	3E	n	6E
S1	0F	?	3F	o	6F
DLE	10	@	40	p	70
DC1	11	A	41	q	71
DC2	12	B	42	r	72
DC3	13	C	43	s	73
DC4	14	D	44	t	74
NAK	15	E	45	u	75
SYN	16	F	46	v	76
ETB	17	G	47	w	77
CAN	18	H	48	x	78
EM	19	I	49	y	79
SUB	1A	J	4A	z	7A
ESC	1B	K	4B	{	7B
FS	1C	L	4C	¦	7C
GS	1D	M	4D	}	7D
RS	1E	N	4E	~	7E
US	1F	O	4F	DEL	7F
SP	20	P	50		
!	21	Q	51		
"	22	R	52		
#	23	S	53		
$	24	T	54		
%	25	U	55		
&	26	V	56		
'	27	W	57		
(	28	X	58		
)	29	Y	59		
*	2A	Z	5A		
+	2B	[	5B		
,	2C	\	5C		
-	2D	]	5D		
.	2E	\	5E		
/	2F	–	5F		

Note Characters hex. 00 to 1F are control characters. Character hex. 7F is delete, or rub-out.

Glossary

Accumulator. A special *CPU register* that receives the results of most *ALU* operations.

A/D Converter. *Analogue* to *digital* converter.

Address Bus. The *microcomputer bus* that carries the *memory* address of the *instruction* that is being fetched, or a *data* item that is being transferred between the *CPU* and memory or *input/output*.

Address Decoder. A circuit that generates *chip select* signals for each *memory* or *input/output chip* within a *microcomputer*.

Addressing Mode. A method of specifying the location of a *data* item which is accessed within an *instruction*.

ALU. Arithmetic and Logic Unit. The module within the *CPU* that performs arithmetic, e.g. add and subtract, and *logic*, e.g. *AND* and *OR* operations.

Analogue. A continuous signal that can take any value over its range.

AND. The *Boolean logic* function that generates logic 1 only if both comparison (or input) *bits* are also at logic 1.

ASCII. American Standard Code for Information Interchange. The code that is used to represent characters in *microcomputers*, printers and *VDUs*.

Assembler. A *program* that converts an *assembly language* program into *machine code*.

Assembly Language. A programming *language* that is line-for-line convertible to *machine code*, but uses *mnemonics* and *labels* to assist the programmer.

Astable Multivibrator. A circuit that generates pulses.

Asynchronous Counter. A *counter* circuit in which each counter stage is triggered after the previous stage (the stages do not trigger at the same time).

Backing Store. A bulk storage device, e.g. *floppy disk* or *hard disk*, for *programs* and *data* files, used by *computers*.

Bandwidth. The width of the band of frequencies passed by a circuit.

Base. The radix of a number system, e.g. the decimal system uses a base of 10 and the *binary* system uses a base of 2.

BASIC. Beginners All-purpose Symbolic Instruction Code. The most popular *high-level language* that is used with *microcomputers*.

Baud Rate. The speed of transmission of *serial data* expressed in *bits* per second.

BCD. Binary Coded Decimal. A *binary* code that uses 4 *bits* to represent the 10 *decimal* numbers.

Binary. A number system that uses the *base* of 2. The only symbols used in binary numbers are 0 and 1. *Computers* represent numbers in binary form.

Bipolar. Possessing charges of opposite electrical polarity. Often used to describe *TTL* circuits in contrast to *MOS* and *CMOS* (unipolar).

Bistable Multivibrator. A circuit element that possesses two stable states. Often known by the name of "flip-flop".

Bit. *Binary* digit. A bit has two states—0 and 1.

Boolean Logic. A collection of *logic* functions named after George Boole. The Boolean logic functions are *AND*, *OR* and *EXCLUSIVE OR* plus complementing functions (*NAND*, *NOR* and *NOT*).

Bounce. Unwanted repeated operation of a mechanical contact.

Breakpoint. A stop that is inserted into a *program* to assist in the testing of a new or faulty program.

Buffer. A temporary storage *register*.

Bug. A *software* error.

Bus. A set of signal connections that have a common function. A *microcomputer* possesses an *address bus*, *data bus* and *control bus*.

Byte. Eight *bits*.

Call. An *instruction* that transfers *program* control to a *subroutine*.

Central Processor Unit. See *CPU*.

Chip. The familiar name for an *integrated circuit*.

Chip Select. A control signal that activates a *memory* or *input/output* chip (normally takes its data connections out of the *floating* state).

Clock. A timing reference for an electronic system.

CMOS. Complementary Metal Oxide Semiconductor. A family of *integrated circuits* that offers extremely high packing density and low power.

Combinational Logic. A *logic gating* system in which all signals change, or are liable to change, together. Contrast with a *sequential logic* system.

Common Mode. An electrical noise signal that is present on both input connections to a circuit.

Comparator. A circuit that compares two signals (*analogue* or *digital*) and indicates the result of the comparison as one of two levels (normally expressed as 0 or 1).

Compiler. A *program* that converts a *high level language* program into *machine code* before program run time and stores both versions of the program on *backing store*. Contrast with an *interpreter*.

Complement. Change a *bit* from 1 to 0 or from 0 to 1.

Computer. A programmable *data* processing system.

Control Register. The addressable *register* on a programmable *input/output chip* that is used to program, or "initialise", the device—found in a *PIO*, *UART* and *CTC*.

Control Unit. The module within the *CPU* that examines and implements the current *instruction*.

Counter. An electronic circuit that implements a count (normally in *binary*) of incoming pulses.

Counter/Timer. A programmable *input/output* circuit that can be used to generate timer *interrupt* pulses, generate time delays or count external pulses. Frequently it is included within a *PIO* chip.

CPU. Central Processor Unit. The main *computer* module, which fetches and implements *program instructions*. Its main sub-modules are the *ALU* and *control unit*. In a *microcomputer* the CPU normally forms a single *chip* and is called a *microprocessor*.

CRO. Cathode Ray Oscilloscope. An item of test equipment that is used to display voltage waveforms on a *CRT* (Cathode Ray Tube).

CRT. Cathode Ray Tube. A display device—used in the domestic television.

Current Tracer. A hand-held fault-finding tool that indicates the magnitude of ac current flowing in an adjacent conductor.

Cursor. A small area of light that appears on the *CRT* screen of a *VDU* and indicates the position at which characters entered by the operator will appear.

D/A Converter. *Digital* to *analogue* converter.

Darlington Driver. An *open-collector driver* circuit that provides a high current drive capability in order to switch electrical devices such as solenoid, relay, motor or lamp.

Data. A general term that can describe numbers, characters or groups of *bits* suitable for processing within a *computer*.

Data Bus. The *microcomputer bus* that carries *data* between *CPU* and *memory* or *input/output*.

Decade Counter. A *binary counter* that counts up to 10 and then resets itself.

Decimal. A number system that uses a *base* of 10.

Decoder. A conversion circuit that activates a single output for a particular coded input.

De Morgan's Rules. Rules of *Boolean* algebra that specify relationships between AND and OR functions.

Denary. An alternative name for *decimal*.

Digit. Each symbol in a number system, e.g. *binary* digit can be 0 or 1.

Digital. Possessing discrete states. Computers operate using *binary* digital signals, i.e. possessing only two states.

DIL. Dual-In-Line. The standard *integrated circuit* package.

Diskette. A small (5¼") *floppy disk*.

DMA. Direct Memory Access. *Data* transfer between *memory* and *input/output* without passing through the *CPU* in a *computer*.

Dot Matrix. A method of constructing characters using an array of dots, e.g. in a printer or *CRT*.

Driver. A circuit that enables a signal to pass to succeeding circuits or long interconnections with reduced electrical deterioration or overloading.

DTL. Diode Transistor Logic. An outdated electronic *logic* family.

D-Type Flip-flop. The most popular *flip-flop* circuit that staticises an input *bit* (on D connection) and passes it through to the output connection (on Q) only when a *clock* pulse is received.

Dual Ramp. The technique adopted by an integrating *A/D* converter.

DVM. Digital Voltmeter. An item of test equipment that displays voltage, current and resistance values in *digital* form.

Dynamic Ram. *RAM* memory that requires a regular refresh operation to prevent corruption of *data*.

ECL. Emitter Coupled Logic. A *bipolar* family of *logic* and gating circuits. Its power consumption is high but its speed is exceptionally fast.

Epitaxial. Part of the fabrication technique that is used to make *planar* circuits.

EPROM. Erasable Programmable Read Only Memory. *ROM* that can be erased by exposure to ultra-violet light and then re-programmed.

Exclusive OR. The *Boolean logic* function that generates logic 1 only if both comparison (or input) bits are different.

Execute. To run a *program*. Alternatively the second part of the *fetch/execute* cycle which is implemented when the *CPU* obeys an *instruction*.

Fan-Out. The maximum number of subsequent *gates* that the output signal connection from a gate can drive.

FET. Field Effect Transistor. The principal component in *MOS* and *CMOS* circuits.

Fetch. The first part in the *fetch/execute* cycle which is implemented when the *CPU* obeys an *instruction*.

Fetch/Execute Cycle. The basic cycle that is implemented by the *CPU* when it obeys an *instruction*. Firstly it is fetched from *memory*, and secondly it is examined by the *control unit* and executed.

Filter. A circuit that passes only a restricted band of frequencies.

Firmware. *Program* or *data* resident in *ROM*.

Flag. A *bit* that indicates a specific condition or event.

Flip-flop. The familiar name for a *bistable multivibrator*.

Floppy Disk. A *backing store* medium that employs flexible magnetic disks.

Flowchart. The diagrammatic representation of the operation of a *program*.

FSK. Frequency Shift Keying. The technique of converting a sinewave signal of a particular frequency to a *logic* level, and vice versa.

Gate. A *digital* circuit with more than one input, but only one output. Gates can perform *Boolean logic* functions.

Glitch. An unwanted pulse or burst of electrical noise.

Hard Disk. A *backing store* medium that employs a non-moveable hard disk. A hard disk is faster, more expensive and possesses larger storage capacity than a *floppy disk*. It is often called a "Winchester".

Hardware. The physical equipment in a *computer*. Contrast with *software*.

Hexadecimal. A number system that uses a *base* of 16. Its particular use is to represent long *binary* numbers, which are used in *microcomputers*, in a shorter form.

High Level Language. A programming *language* that is similar to spoken language. A high level language *program* must be converted into *machine code* before it is run in a *computer*.

Hysteresis. The effect of lag in a circuit whereby the output does not appear fully to respond to the activating signal.

IBM 3740. An industry standard for the format of *data* on a *floppy disk*.

IC. *Integrated circuit.*

Initialise. To set an *input/output chip*, e.g. a *PIO, UART* or *CTC*, to one of its programmable states.

Input/Output. The *hardware* within a *computer* that connects the computer to external *peripherals* and devices.

Input Port. A circuit that passes external *digital* signals (normally 8) into a *microcomputer*.

Instruction. A single operation performed by a *computer*. A *low level language program* consists of a list of instructions.

Integrated Circuit. A circuit package that contains several components built into the same semiconductor wafer.

Interface. To interconnect a *computer* or other electronic system to external devices and circuits.

Interpreter. A *program* that converts a *high level language* program into *machine code* at run-time, and then executes that sequence of machine code instructions for each high level language command. Contrast with a *compiler*.

Interrupt. An external signal that suspends a *program* operating within a *computer* and causes entry into a special interrupt program.

Invert. To change the polarity of a signal.

J–K Flip-flop. A *flip-flop* circuit that has two inputs and a *clock* signal. Its particular feature is that its output "toggles" (switches from one state to the other) on succeeding clock pulses if its two inputs are held at 1. It is used in *counters*.

Jump. An *instruction* that sends *program* control to a specified *memory* location.

K. A symbol that represents *decimal* 1024.

Kansas Standard. A signal specification for *data* storage on audio cassette recorders.

Label. A name given to a *memory* location in an *assembly language program*.

Language. A prescribed set of characters and symbols which is used to convey a *program* to a *computer*. A programming language can be a *high level language* or a *low level language*.

Latch. A circuit that staticises *bits*.

LCD. Liquid Crystal Display. A very low-powered numeric display that operates on the principle of reflecting incident light.

LED. Light Emitting Diode. A diode that emits light when current passes through it. A LED is commonly applied with *microcomputers* to indicate a condition or event, e.g. machine on/off, and in the structure of *segment displays*.

Live Zero. A signal range that does not possess a zero quantity, e.g. 4–20 mA.

Logic. The application of a range of circuit building blocks to perform switching and control functions.

Logic Analyser. An item of test equipment that can display the states of several *logic* levels.

Logic Level. The voltage value that is used to indicate *logic* 0 or 1. Normally +5 V = logic 1, and 0 V = logic 0.

Logic Probe. A hand-held item of test equipment that uses a *LED* (or LEDs) indicator to highlight the logic state of any circuit test point.

Logic Pulser. A hand-held item of test equipment that is used to inject pulses into a circuit under test.

Loop. A section of *program* that is executed more than once.

Low Level Language. A *computer* programming *language* that specifies each operation that the *CPU* is to perform. There are two classifications of low level language: *assembly language* and *machine code*.

LSI. Large Scale Integration. A measure of the packing density of an *integrated circuit* (greater than 100 *gates* per *chip*). See *SSI, MSI* and *VLSI*.

M. A symbol that represents a million. See also *K*.

Machine Code. A program expressed in *binary* form, i.e. in the way in which it is loaded into *memory* and executed within the *CPU*.

Main Memory. Fast *memory* which holds the *program* currently being executed. Main memory can be *ROM*, *RAM* or a mixture of the two.

Master-slave Flip-flop. A *flip-flop* in which the input is not transferred immediately to the output. A complete *clock* pulse (rising edge plus falling edge) is necessary to switch the flip-flop. Other flip-flops are edge (or level) triggered.

Matrix. An array of intersecting rows and columns with logic components (e.g. pushbutton keys) connected at the intersections.

Matrix Printer. A printer that constructs characters using a *dot matrix*.

Memory. Any circuit or *peripheral* that staticises *data*. Normally the term is used in place of *main memory* in a computer.

Memory Map. A diagrammatic representation of the address assignments within a *computer*.

Memory Mapped Input/Output. *Input/output* devices that are treated by *hardware* and *software* as *memory* devices.

Microcomputer. A complete *computer* on a handful of *integrated circuits* (or even a single integrated circuit). *VLSI* components are used for *CPU*, *memory* and *input/output*.

Microelectronics. The construction of electronic circuits in microminiaturised form.

Microprocessor. A *CPU* constructed on a single *VLSI integrated circuit*.

Mnemonic. A group of letters that is used to represent the function of an *instruction* expressed in *assembly language* form.

Modem. An item of equipment that converts *logic* levels to frequencies, and vice versa. A modem (abbreviation for modulator–demodulator) is used when it is required to connect a *serial data* link through the telephone network.

Monitor. The main *program* in many *microcomputers*.

Monostable multivibrator. A two-state circuit that has only one stable state.

MOS. Metal Oxide Semiconductor. A family of *integrated circuits* that offers high packing density. MOS technology is used to construct *microprocessors*, *ROM*, *RAM* and many *input/output* devices.

MSI. Medium Scale Integration. A measure of the packing density of an *integrated circuit* (greater than 10 *gates* per *chip*). See *SSI, LSI* and *VLSI.*
MTBF. Mean Time Between Failures.
MTTR. Mean Time To Repair.
Multimeter. An item of test equipment that measures several electrical parameters over several different ranges, e.g. a *DVM.*
Multiplexing. The technique of passing more than one signal along a single conductor.
Multivibrator. A circuit that can be set into one of two states—see *astable multivibrator, bistable multivibrator* and *monostable multivibrator.*

NAND. The *Boolean logic NOT AND* function.
Negative Logic. The representation of *logic* 1 by a low voltage and logic 0 by a high voltage.
Nesting. A *program loop* within another loop. A *subroutine* within another subroutine.
Node. A point in a circuit at which two or more signals join.
Noise Immunity. A measure of the ability of a circuit to reject noise.
NOR. The *Boolean logic NOT OR* function.
NOT. The *Boolean logic* function that inverts a logic level or a *bit.*
Nybble. Four *bits* ($\frac{1}{2}$ *byte*).

Object Code. The name given to a *machine code* version of a *program.* The term is used to distinguish this version from the *assembly language* version (called the "source program").
One's Complement. The complement of a *binary* number. It is created when the *two's complement* version of a binary number is being generated.
Op-Amp. Abbreviation for Operational Amplifier.
Opcode. The part of a *machine code instruction* that specifies the function of the instruction, e.g. add, *shift, jump.*
Open-Collector Driver. A *TTL* circuit in which the output signal requires an external load resistor—the top half of the "totem pole" circuit is absent.
Operand. The part of a *machine code instruction* that specifies the *data* value or its *memory* address.
Operating System. The main *program* in a *disk*-based *microcomputer.*
Operational Amplifier. An amplifier for *analogue* signals that amplifiers dc as well as ac signals.
OR. The *Boolean logic* function that generates logic 1 if either of the comparison (or input) *bits* is set to logic 1.
Oscillator. A circuit that generates sinewaves.
Output Port. A circuit that passes *digital* signals (normally a group of 8) outside a *microcomputer.*

Parity. The number, expressed as odd or even, of 1s in a *data* value.
PCB. Printed Circuit Board. A conventional circuit board with etched copper track interconnections between components.
Peripheral. An item of equipment that is external to a *computer*, e.g. printer, *VDU, floppy disk.*
Phase Locked Loop. A circuit that generates a predetermined *logic* level at its output only if a precise frequency signal is present at its input.
Piggy-backing. The technique of placing an *IC* on top of another identical IC.
PIO. Parallel Input/Output. A programmable multi-*port input/output chip.*
Pixel. A dot position on a *CRT* screen that is driven by a *computer.*
Planar. The fabrication process (photo-mask and diffusion stages) that is used in the fabrication of the great majority of *integrated circuits.*
Poll. To regularly check the status of an external signal or device by *software.*
Port. An input or output channel between a *microcomputer* and external equipment. Normally a port is 8-*bits* wide.

Positive Logic. The representation of *logic* 1 by a high voltage and logic 0 by a low voltage.

Program. A set of processing steps that a *computer* is required to perform.

Program Counter. A *CPU register* that holds the address in *memory* of the next *instruction* to
· be obeyed.

PROM. Programmable Read Only Memory. *ROM* that is programmed after the *chip* is
manufactured. Once programmed it cannot be altered.

Propagation Delay. The time for a *logic* level change to propagate through a circuit.

Pseudo-Instruction. An *instruction* in an *assembly language program* that does not cause the
generation of a *machine code* instruction when the program is assembled.

Pulse Generator. A circuit that generates pulses—see *astable multivibrator*.

RAM. Random Access Memory. RAM is semiconductor read/write *memory*.

Read. To transfer *data* from *memory* to the *CPU* in a *computer*.

Refresh. To re-instate *data* stored in *dynamic RAM* or displayed on a *CRT* or *segment display*.

Register. A storage device for several *bits*. A *microprocessor* contains several work registers,
which can be used for temporary storage of *data* within a *program*.

Resistor Ladder. The technique adopted by the most common form of *D/A converter*.

Return. An *instruction* that returns *program* control to a main program from a *subroutine* or an
interrupt program.

Ripple-through Counter. A *counter* in which the output of one stage is connected as the input
to the succeeding stage.

ROM. Read Only Memory. ROM is semiconductor *memory* which can only be read. There are
three common classifications of ROM: ROM, *PROM* and *EPROM*.

RS232-C. The internationally-recognised specification for *serial data* transfer between *computers* and serial-drive *peripherals*.

Schmitt Trigger. A circuit that is applied to generate a well-defined pulse shape from an input
voltage waveform that features slow or irregular rise and fall characteristics. The circuit
possesses a large degree of "*hysteresis*"".

Segment Display. A display that constructs numbers and letters by a network of segments.
Microcomputers commonly use 7-segment displays for representing numbers.

Semiconductor Memory. *ROM* and *RAM*.

Sequential Logic. A *gating* and *logic* system that employs memory elements, and in which
signals change at various points in the system only as a result of signals at previous points in
the system changing.

Serial. The transfer of *data* items by setting one *bit* at a time on a single conductor.

Shift. Transfer of *binary data* to the left or right.

Shift register. A *register* in which the stored *data* can be shifted to the left or right.

Shotgunning. The procedure of replacing each *IC* in turn during a fault-finding exercise until
the fault is cleared.

Signature Analyser. An item of test equipment that is applied with *microcomputers*. A
4-character hexadecimal "signature" is displayed for different circuit test points.

Silicon. The semiconductor element that is applied to fabricate virtually all *integrated circuits*.

Sink. To accept current from the succeeding circuit.

Software. *Computer programs* and *data* files.

Source. To provide current flow into the succeeding circuit.

Source program. The name given to the *assembly language* version of a *program*.

SSI. Small Scale Integration. A measure of the packing density of an *integrated circuit* (less than
10 *gates* per *chip*). See *MSI, LSI* and *VLSI*.

S–R Flip-flop. The most basic type of *flip-flop*, which does not possess a *clock* input. It possesses
the disadvantage of having an indeterminate state when its two inputs are set together.
(S = Set, R = Reset).

Stack. A reserved area of *memory* (*RAM*) that is used in many *microprocessors* to store the return address in *subroutines* and *interrupt programs*.

Status Register. A collection of *flag bits* in a *microprocessor* that indicates the state of the *ALU*.

Strobe. A signal that is used as a reference in an electronic system. The term is also used to describe the action of reading an external signal (or signals) into a *CPU* under *software* control.

Subroutine. A section of *program* that is separated from the main program, but can be called several times from the main program.

Successive Approximation. The technique of generating successive fractions of a reference voltage in an *A/D converter*.

Synchronous Counter. A *counter* circuit in which all counter stages are triggered together. Contrast with an *asynchronous counter*.

Telemetry. The technique of transmitting a large number of *digital* signals over a large distance using *multiplexing*.

Terminal. An operator station at which *data* entry is made to a *computer*. A terminal is normally a *VDU*, although sometimes it is a printer keyboard.

Three-State (or Tri-State). A circuit in which its outputs can be set into one of three states—*logic* 0, logic 1 or "floating" (high impedance state, i.e. electrically disconnected).

Toggle. The reversal of a *binary* signal.

Totem Pole. The output stage of a normal *TTL* circuit.

Transducer. A component or device that converts one form of energy to another, e.g. temperature to electrical.

TTL. Transistor Transistor Logic. A family of *integrated circuits* that preceded *MOS* and *CMOS*, but still offers the advantage of much faster operating speed. However packing density is low in TTL and power consumption is much higher than for MOS and CMOS. Typically a full *microcomputer* circuit includes MOS *CPU*, *ROM* and *RAM*, but also TTL *gates* and *buffers*.

TTL Compatible. The characteristic of a circuit whereby its input and output signals can be connected to *TTL* devices.

TTY. Teletype. A name sometimes given to a printer.

Two's Complement. A *binary* numbering system used to represent both positive and negative numbers.

UART. Universal Asynchronous Receiver Transmitter. An *input/output chip* that handles *serial data* transfer, e.g. to *VDU*, printer or other *computer*.

Unipolar. A transistor that uses only one charge carrier. A more familiar name is *FET*.

VDU. Visual Display Unit. An operator device that includes a *CRT* for display purposes and a keyboard for manual entry.

Vector. A fixed *memory* location that contains the start address of an *interrupt program*.

Virtual Earth. A principle of circuit analysis that is applied with *operational amplifiers* (it is assumed that the input current is zero).

Volatile Memory. *Memory* that loses its stored *bit* pattern when power is removed, e.g. *dynamic RAM*.

VLSI. Very Large Scale Integration. A measure of the packing density of an *integrated circuit* (more than 1000 *gates* per *chip*). See *MSI*, *LSI* and *SSI*.

Wafer. A slice of *silicon* crystal into which *planar* circuits are fabricated during the manufacture of *integrated circuits*.

Winchester. Another name for a *hard disk.*

Wire-AND. A circuit arrangement that uses *TTL open-collector* devices with their outputs connected together to generate a *logic AND* function.

Word. A unit of *data* in a *computer.* The word length is the same as the *bit* length of the *microprocessor,* i.e. normally either 8 bits or 16 bits.

Write. To transfer *data* from *CPU* to *memory.*

Write Protect. To set a *backing store* device to read-only to protect against over-writing.

Zap. To change the contents of a *computer's memory* location.

ZIF Socket. Zero Insertion Force socket. A socket into which *integrated circuits* in *DIL* form can be inserted and then clamped by means of a locking lever.

Answers to Exercises (Numerical)

Chapter 2
1. 0 V
2. 14
3. 0
4. A
7. $X = \overline{C + (A + B)}$ and 1

Chapter 3
4. 10

Chapter 4
11. A = 0, B = 1, C = 0, D = 0

Chapter 5
1. $R_B = 10\,k\Omega$, $R_A = 1\,M\Omega$
4. Logic 0
5. 101001
8. $(A + C).(B + C)$

Chapter 6
1. 101001
2. 92
3. 8D
4. 1010 0110
5. 46
6. 45
7. 0100 1101
8. 0101 0101
9. 5B
10. 0D

Chapter 7
4. $R_1 = 100\,k\Omega$, $R_f = 10\,k\Omega$
5. 5.7

Chapter 8
5. 2.7 V

Chapter 9
1. 1 in 4096 (0.25%)

Chapter 11
8. RAM—1000, 13FF
   EPROM—2000, 27FF
14. C000

Chapter 12
6. 960
10. 163,840

Subject Index